Disclaimer

The publisher of this book is by no way associated with the National Institute of Standards and Technology (NIST). The NIST did not publish this book. It was published by 50 page publications under the public domain license.

50 Page Publications.

Book Title: 2005 Programs of the Manufacturing Engineering Laboratory

Book Author: Lisa J. Fronczek; Bessmarie A. Young

Book Abstract: The National Institute of Standards and Technology?s Manufacturing Engineering Laboratory (MEL)strengthens the U.S. economy and improves the quality of life by working with the U.S. manufacturing industry to develop and apply infrastructural technology, measurements, and standards to meet their needs. This report contains summaries of MEL programs that support the needs of the U.S. manufacturing industry. Each program summarizes the resources, objectives, customer needs that are addressed, accomplishments, current year plans, lifetime objectives, and related measurement and standards work.

Citation: NIST Interagency/Internal Report (NISTIR)

Keyword: Manufacturing, manufacturing engineering,technology, measurements, metrology, standards

MEL
innovation & productivity

NIST
National Institute of Standards and Technology
Technology Administration, U.S. Department of Commerce

Dale Hall
Director, dhall@nist.gov
Howard Harary
Deputy, hharary@nist.gov

Precision Engineering
Dennis Swyt, Chief
dswyt@nist.gov

Large Scale Coordinate Metrology
Steven Phillips, sphillip@nist.gov

Engineering Metrology
Ted Doiron, tdoiron@nist.gov

Nanometer-Scale Metrology
Michael Postek, mpostek@nist.gov

Surface and Microform Metrology
Theodore Vorburger, tvorburger@nist.gov

Manufacturing Metrology
Kevin Jurrens, Chief (acting)
kjurrens@nist.gov

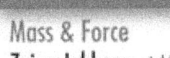

Mass & Force
Zeina Jabbour, zjabbour@nist.gov

Machine Tool Metrology
Alkan Donmez, adonmez@nist.gov

Sensor Development & Application
Kang Lee, kang.lee@nist.gov

Mfg. Process Metrology
Robert Polvani, rpolvani@nist.gov

Manufacturing Systems Integration
Steven Ray, Chief
sray@nist.gov

Deputy
Sharon Kemmerer, skemmerer@nist.gov

SIMA Office
Simon Frechette, sfrechette@nist.gov

Enterprise Systems
Albert Jones, ajones@nist.gov

Mfg. Standards Metrology
Simon Frechette, sfrechette@nist.gov

Design & Process
Ram Sriram, rsriram@nist.gov

Mfg. Simulation & Modeling
Chuck McLean, cmclean@nist.gov

Intelligent Systems
Al Wavering, Chief
awavering@nist.gov

NIST Fellow
James Albus, jalbus@nist.gov

Control Systems
Frederick Proctor, fproctor@nist.gov

Knowledge Systems
Elena Messina, emessina@nist.gov

Machine Systems
Al Wavering, awavering@nist.gov

Perception Systems
Michael Shneier, shneier@nist.gov

Systems Integration
Maris Juberts, mjuberts@nist.gov

Fabrication Technology
Mark Luce, Chief
mluce@nist.gov

Main Shops
Bob Wantz, Asst., rwantz@nist.gov

MEL
innovation & productivity

the programs 2005 of the manufacturing engineering laboratory

NISTIR 7218

January 2005
Revised April 2005

Abstract

The National Institute of Standards and Technology's Manufacturing Engineering Laboratory (MEL) strengthens the U.S. economy and improves the quality of life by working with the U.S. manufacturing industry to develop and apply infrastructural technology, measurements, and standards to meet their needs. This report contains summaries of MEL programs that support the needs of the U.S. manufacturing industry. Each program summarizes the resources, objectives, customer needs that are addressed, accomplishments, current year plans, lifetime objectives, and related measurement and standards work.

Keywords

Manufacturing, manufacturing engineering, technology, measurements, metrology, standards

Disclaimer

Certain commercial equipment, instruments, or materials are identified in this report in order to specify the experimental procedure adequately. Such identification is not intended to imply recommendations or endorsement by the National Institute of Standards and Technology, nor is it intended to imply that the materials or equipment identified are necessarily the best available for the purpose.

For More Information

Manufacturing Engineering Laboratory
Stop 8200
Gaithersburg, MD 20899-8200

Phone 301-975-3400
www.mel.nist.gov

Table of Contents

Director's Message	1
Introduction / User's Guide	3
MEL Capabilities	7
MEL Facilities	11

Program Descriptions

	At a Glance	Full Write-up
Program-Division Cross-Reference Chart		28
Introduction to Programs		29
Dimensional Metrology	19	31
Homeland and Industrial Control Security	20	44
Intelligent Control of Mobility Systems	21	54
Manufacturing Interoperability	22	66
Manufacturing Metrology and Standards for the Health Care Enterprise	23	79
Mechanical Metrology	24	86
Nanomanufacturing	25	98
Smart Machining Systems	26	111

Special Activities

Introduction to Special Activities	124
Competence Projects	126
Exploratory Projects	134
Recently Completed Exploratory Projects	149
The National Science and Technology Council Interagency Working Group on Manufacturing Research and Development	159
Intelligent Manufacturing Systems (IMS)	162
Systems Integration for Manufacturing Applications (SIMA)	166

Director's Message

Dale Hall, 301 975 3400, dale.hall@nist.gov

I am proud to present the technical work of the Manufacturing Engineering Laboratory (MEL). The work described in this book shows how MEL addresses critical measurements, standards, and advanced technology needs of our manufacturing customers. It also describes how MEL work contributes to the development of the measurements and standards infrastructure needed for the emerging technology areas of the future. Our new FY 2005 programs are described, as well as the many significant accomplishments of FY 2004. Again this year, we incorporated suggestions from our customers to make this book a more useful source of information on our laboratory and its programs.

Many of our most demanding metrology projects and services moved into the new NIST Advanced Measurements Laboratory in 2004. This new $235 million facility gives us an unparalleled environment for state-of-the-art dimensional and mechanical metrology and ensures that we will continue to provide our customers with the best available measurement technology.

The needs of our customers are embedded in our programs from their onset; we remain committed to providing customer value as well as the highest technical quality.

In 2004, MEL carried out an intensive strategic planning exercise to ensure the future vitality, quality, and productivity of the laboratory's technical programs. Our new portfolio of eight technical programs uses our resources more effectively and focuses our work on the critical technologies and measurement services our customers need. I also believe that our technical priorities and objectives are now sharper and easier for our customers and stakeholders to understand.

In 2004, we assessed our current and projected resource needs and resource projections and developed an MEL resource acquisition strategy. In 2005, we intend to use this strategy to pursue resources that will support work that is well aligned with our mission, technical capabilities, and interests. As part of that effort, we are working hard to improve our marketing skills and practices.

Programs of the Manufacturing Engineering Laboratory

I am excited about the new role that we play in the newly formed Interagency Working Group (IWG) on Manufacturing Research and Development (R&D). This group, of which I am vice-chair, will continue the work of the Government Agencies Technology Exchange in Manufacturing (GATE-M) to improve cooperation and coordination among federal agencies involved in manufacturing R&D. This year, the IWG for Manufacturing R&D identified three technical priority areas: Intelligent and Integrated Manufacturing Systems; Manufacturing for the Hydrogen Economy; and Nanomanufacturing. In each of these areas, the IWG commissioned white papers to outline the challenges and technical progress needed as a basis for future federal R&D efforts. A public forum for comment on the IWG technical priorities is planned for early 2005.

In April 2005, the Intelligent Manufacturing Systems (IMS) program, an international effort to address manufacturing challenges and sustainable production systems in the 21st century, will complete a ten-year Phase I effort. For most of that time, the U.S. Secretariat of IMS has been located in MEL. The IMS International Steering Committee is now planning Phase II. I am looking forward to working with the new head of the U.S. Delegation, who intends to increase U.S. corporate participation to enhance the value of this program to U.S. manufacturing.

In its current strategic plan, NIST identified four areas for future technical focus: Nanotechnology, Information/Knowledge Management, Homeland Security, and Biosystems and Health. NIST has created or is forming Strategic Working Groups in each of these areas. A superbly qualified MEL staff member leads the Nanotechnology Strategic Working Group; MEL staff members serve on the other strategic working groups as well. MEL also continues to promote manufacturing as a high priority at NIST and is working proactively to improve the coordination of NIST's manufacturing effort and to promote a more assertive external presence in manufacturing technology. NIST has a broad array of programs to help manufacturers, and other federal agencies look to NIST to take a leadership role.

I'm very proud of the value we provide to our customers, the superb technical quality of MEL's work, and especially the creativity and dedication of our staff. MEL is a vital asset for the future of US manufacturing. All of us in MEL look forward to working with and serving you – our customer. We're always pleased to receive your comments and your questions.

Programs of the Manufacturing Engineering Laboratory

Introduction/ User's Guide

Introduction/ User's Guide

The National Institute of Standards and Technology (NIST) Manufacturing Engineering Laboratory (MEL) is pleased to present a summary of its technical programs for fiscal year (FY) 2005. MEL's mission is to satisfy the measurements and standards needs of U.S. manufacturers in mechanical and dimensional metrology and advanced manufacturing technology. We do this by performing research and development (R&D), providing services, and participating in standards activities. Our long-term goal is best summed up by the Laboratory's core purpose — to promote a healthy U.S. manufacturing economy by solving tomorrow's measurement and standards problems today.

With a value-added contribution of $1.4 trillion, U.S. manufacturing — our customer base — directly accounts for approximately 13 percent of the U.S. gross domestic product (GDP). Manufacturing plays a central role to our Nation's economy. Innovation seen in the manufacturing industry is vital to all the other sectors of the economy.

We serve the manufacturing sector of the U.S. economy in a broad sense, working with partners from industry as well as other government agencies and academia to develop the measurement tools and infrastructure that enable higher productivity, new products, and improved processes. Our work primarily supports the manufacture of durable goods, including discrete, mechanical, and electromechanical products. We also exploit opportunities to apply our technology to support other key industrial sectors, including the manufacture of non-durable goods and non-discrete products. Our customers span the full range from established to emerging-technology industries, including automotive, aerospace, construction equipment, electronics, optics, telecommunications, and nanotechnology. MEL also provides design and fabrication services to other NIST operating units through its Fabrication Technology Division.

MEL actively develops and maintains strong relationships with its customers and stakeholders. In fact, many of the nation's leading manufacturers rely on MEL; therefore, we develop and operate programs in direct response to their needs. Working collaboratively with our partners, our staff members solve measurement and standards problems that allow our industrial customers to take full advantage of technology such as advanced manufacturing techniques and on-line quality assurance processes. Our customers depend on MEL for calibration services that are the best in the world. These services, for example, ensure dimensional compatibility of items manufactured at different sites and satisfy requirements for traceability to national standards. MEL maintains an active program of technology transfer through cooperative research, industrial research associates who come to our laboratories, publications, conferences, seminars, workshops, and customer visits. MEL's interactions extend from the bench researcher to senior management.

Programs of the Manufacturing Engineering Laboratory

MEL staff continued to serve our customers in other important ways. The measurement methods, calibration services, software interface standards, and leading-edge manufacturing techniques all support U.S. industries' efforts to improve performance and provide quality products in the laboratory, the design shop, the factory, and ultimately, in the marketplace.

Our technical work is carried out through strategic programs as well as division or special projects. This document summarizes those efforts and is organized as follows.

MEL Capabilities and Facilities
(starting on page 7)

MEL's staff expertise is diverse, a reflection of the multi-disciplinary manufacturing measurement and standards problems that MEL addresses. The staff's individual expertise, coupled with MEL's outstanding and often unique facilities result in a rich array of technical capabilities that can address manufacturing problems spanning the entire manufacturing enterprise, both virtual and physical, from design to the factory floor and beyond.

Program-at-a-Glance
(starting on page 17)

These one-page, high-level descriptions provide a quick overview of each technical program, including the goal, a problem statement, technical approach, and typical customers and collaborators.

MEL Program-Division Cross-Reference Chart
(page 28)

MEL is comprised of five divisions that are responsible for specific mission areas that address the complexity and demands of the manufacturing community:

- Precision Engineering Division provides the foundation for dimensional measurements over 12 orders of magnitude (hundreds of meter to nanometers);

- Manufacturing Metrology Division fulfills the measurements and standards needs in mechanical metrology and advanced manufacturing;

- Intelligent Systems Division focuses on the measurement, standards, development and application of intelligent control, open-architecture standards, and intelligent systems manufacturing;

- Manufacturing Systems Integration Division promotes U.S. economic growth by working with industry to develop and apply measurements and standards that advance information-based manufacturing technology; and

- Fabrication Technology Division provides world-class instrument and specialized fabrication support for NIST researchers and serves as a testbed for many NIST and MEL programs.

The divisions are organizational units that are ultimately responsible for long-term responsibilities (e.g., maintaining specialized equipment, meeting staffing needs, and developing and maintaining core technical competencies.) While each division has its own focus, collaborations between divisions are used to achieve MEL strategic program goals.

Programs of the Manufacturing Engineering Laboratory

Program Descriptions
(starting on page 29)

Traditionally, MEL's programs focused on measurement and standards for making things traceable, right, small, and interoperable. This year, in addition to those areas, two new programs focus on measurement and standards for making things safe and healthy. Our programs address advanced manufacturing research and measurement services in dimensional and mechanical metrology, manufacturing processes and equipment, systems integration and interoperability, and intelligent controls.

Within the program's 5-year planning horizon, milestones or technical objectives are set to help achieve the program goal. This section provides comprehensive descriptions of each of MEL's 8 FY 2005 programs, including its goal, customer needs, technical approach, program objectives, technical outputs, and significant accomplishments.

Special Activities
(starting on page 123)

In addition to their work on MEL strategic programs, MEL staff work on a variety of other activities such as competence projects, exploratory projects, coordinate manufacturing-related R&D throughout NIST, participate in manufacturing R&D at the federal level, and participate in an international program designed to increase U.S. manufacturing competitiveness.

Appendix
(starting on page 171)

This section provides a list of commonly used acronyms for quick reference.

MEL Capabilities

MEL Capabilities

MEL's expertise is diverse, a reflection of the multi-disciplinary manufacturing measurement and standards problems that MEL addresses. MEL's staff includes experts in electrical and mechanical engineering, physics, mechanics, materials, acoustics, chemistry, mathematics, computer science, operations research, and machining. These varied backgrounds, coupled with MEL's outstanding and often unique facilities, comprise a rich array of technical capabilities that can address manufacturing problems spanning the entire manufacturing enterprise, both virtual and physical, from design to the factory floor and beyond.

The diverse problems addressed by MEL also span a variety of manufacturing industries. MEL performs dimensional measurements on objects ranging from ship hulls to atomic step heights. MEL measures forces ranging from that generated by spacecraft to that required to break covalent bonds. MEL's intelligent system work has been applied to automated systems ranging from Search and Rescue robots to self-driving Army vehicles. MEL's interoperability data work has application in both manufacturing design and in patient records for the health care industry. These are but a few examples.

MEL's capabilities can be grouped into six different areas:

In precision engineering, MEL realizes and disseminates the SI (Système International d'unités, or International System of units) unit of length, thus providing the foundation of dimensional measurement that meets the needs of the U.S. industrial and scientific communities. MEL delivers to industry important length-related measurements, standards, and technology services that support U.S. manufacturing's products and processes. MEL is capable of measuring feature dimensions spanning over 12 orders of magnitude, from hundreds of meters down to less than a nanometer. MEL can also measure form and surface roughness using a variety of instruments and devices, including coordinate measuring machines; frameless coordinate length-metrology systems involving mechanical-probe, laser-ranging, theodolite, and related interferometric systems; tunneling microscopes; mechanical profilers; phase-measuring interferometers; and tunneling, atomic-force, electron, and visible- and ultraviolet-light microscopes.

Manufacturing Engineering Laboratory Capabilities

In mechanical metrology, MEL provides a similar foundation for meeting the needs of the U.S. industrial and scientific communities by realizing and disseminating the SI mechanical units of mass, force, acceleration, and sound pressure. MEL activities include realizing, maintaining, and improving the primary standards for these quantities; conducting intercomparisons to coordinate and establish comparability of these standards to those of other countries; developing suitable mechanisms to enable the transfer of accurate measurement capabilities to our customers; and providing efficient test and calibration services of the highest quality. MEL also works to extend and improve the mechanical measurements, standards, and technology services by conducting on-going research and development that can lead to new measurement services, new ranges of measurements (e.g., micro-Newton forces), or new realizations of units (e.g., Watt-based kilogram). Several specialized instruments and facilities contribute to the unique MEL mechanical metrology capabilities, including the large acoustic anechoic chamber, force metrology 4.44 MN deadweight machine, and U.S. National Prototype Kilogram mass metrology laboratory.

In manufacturing metrology MEL develops methods, models, sensors, and data to improve metrology, machines, and processes to meet the needs of U.S. manufacturers. To provide the knowledge base needed for efficient development of manufacturing methods MEL explores advanced manufacturing processes and machines to support U.S. industry's need to improve productivity, tighten tolerances, and machine new materials. Areas of work include high-speed machining, high strain rate materials properties, machining process metrology and optimization, advanced optical fabrication and metrology, metrology for machining meso-scale parts, smart sensors for machine/process monitoring and control, and machine tool and subsystem performance characterization.

MEL also develops the measurements and standards infrastructure needed for the application of intelligent systems by manufacturing industries. MEL conducts research to develop a basic understanding of the theoretical issues of intelligent systems, including architecture frameworks, sensory perception, and knowledge representation. MEL develops tools and engineering methodologies for building real-time controls for intelligent systems. MEL applies this research and these tools and methodologies to assist industry in the development, implementation, and testing of standards and performance measurement techniques to enable the effective, efficient, safe, and secure use and diffusion of intelligent systems.

Manufacturing Engineering Laboratory Capabilities

MEL provides high-quality, reliable, and cost-effective fabrication and technical support services to NIST staff. This work includes assisting in the design and construction of instruments and devices needed to maintain the national and international standards of measurement and measurement services for which NIST has stewardship. MEL provides the engineering and manufacturing expertise and equipment needed to produce specialized parts and subassemblies for one-of-a-kind instruments conceived by NIST scientists and engineers.

In manufacturing systems integration MEL works with industry to develop and apply measurements and standards that advance the use of information-based manufacturing technology. MEL engages in work at several levels of abstraction in system integration capabilities: measurements and standards, process representation, integration, modeling capabilities, and the use of software to enhance manufacturing performance. The work supports the foundational areas of interoperability and metrology.

MEL Facilities

MEL Facilities

Force 4.44 MN Deadweight machine

MEL has a number of specialized facilities, many of which are unique in the world, support our programmatic work and enable our high-precision measurement services. These facilities allow MEL to realize technical accomplishments that are not obtainable at other facilities. They also allow for the provision of measurements to a level of accuracy and precision unrealizable by other providers. Collaborators and customers come to take advantage of such facilities.

MEL's specialized facilities include:

Acoustic Anechoic Chamber
This vibration-isolated, shell-within-shell structure is one of the quietest and best acoustically characterized rooms in the world. Typical expanded uncertainties for pressure calibrations of customer microphones in this chamber are less than 0.25 dB over a frequency range of 50 Hz to 17 kHz.

Force 4.44 MN Deadweight machine
The MEL 4.44 Mega-Newton (MN) deadweight machine is the largest in the world, and has an impressive expanded measurement uncertainty of five-parts-per-million. All other large deadweight machines in the world are traceable to this one.

Mass Standards Facility
This is a class 1000 clean room with 0.1 °C temperature control and 2 % relative humidity control, with mass comparators that enable kilogram comparisons to the national prototype kilogram to a combined standard uncertainty of less than 20 μg and ten-kilogram comparisons to a combined standard uncertainty of less than 200 μg.

Manufacturing Engineering Laboratory Facilities

Microforce Realization and Measurement Laboratory

This world-class facility provides a force measurement capability traceable to the SI system of units for forces as small as 10^8 N – a force roughly equivalent to that necessary to break three covalent bonds. Using a null-force electromechanical balance, traceable measurement of small forces can be performed, eventually on the scale of 10^{-12} N (molecular scale forces), based on precision measurements of capacitance, voltage, and length.

Pulse-Heated Kolsky Bar Facility

This facility provides a unique capability to measure stress-strain relationships of materials under high heating-rate, high strain-rate conditions. This one-of-a-kind measurement system can produce mechanical strain rates of 10^4 s^{-1} to 10^6 s^{-1}, and heating rates of 10^4 $°C$ s^{-1} within test materials. The resulting material properties enable improved predictive simulations for such applications as high-speed machining and armor protection.

Smart and Wireless Sensors (SAWS) Laboratory

This facility is used to develop, evaluate, and demonstrate sensor networking and sensor interoperability technologies and standards used for applications ranging from machine tool condition monitoring, military command and control, first responder communications, and on-board monitoring of ship drive systems.

Molecular Measuring Machine (M^3)

Reference Test Arena for USAR/EOD Robots

MEL developed and built this portable 400 m^2 test course to evaluate and promote the development of sensing, navigation, and mapping capabilities of mobile robots, particularly those used in urban search and rescue (USAR) and explosive ordinance disposal (EOD).

Large Scale Coordinate Metrology Laboratory

This facility is used to characterize measurement systems and evaluate metrology techniques needed to measure large-scale objects accurately.

Molecular Measuring Machine (M^3)

The M^3 is an experimental two-dimensional measuring machine capable of measuring the placement of sub-nanometer features with sub-nanometer resolution within a 50 mm- by 50 mm-sized scanning area.

Manufacturing Engineering Laboratory Facilities

Precision Machining Research Facility
This facility is used for developing and validating improved metrology instrumentation and procedures for characterizing machine tools, metal cutting, and other advanced manufacturing processes.

XCALIBIR – eXtremely accurate optics CALIBration InterferometeR
XCALIBIR is a world-class instrument for measuring the surface figure errors of precision optics with a measurement uncertainty of 0.25 nm for flats, spheres, and mild aspheres with a nominal aperture of 300 mm.

Intelligent Mobility Testbed
This testbed consists of autonomous off-road vehicles (i.e., two specially instrumented passenger vehicles and a chase vehicle) and characterized terrain and test courses to advance the state-of-the-art for intelligent vehicles.

Industrial Control System Networking and Security Testbed
This testbed includes a representative array of industrial process automation and advanced networking equipment used to 1) validate standards, and 2) develop and evaluate test methods.

M48 Coordinate Measuring Machine (CMM)
The M48 CMM is a three-dimensional mechanical-contact and vision-based measuring machine for the calibration of 1-, 2-, and 3-D reference standards, with dimensions up to 1 meter, at sub-micrometer accuracies.

The NIST Shops (MEL Fabrication Technology Division)
The Shops provide specialized fabrication and technical support services to NIST researchers, including:
- Optical fabrication
- Glassblowing and glass fabrication
- CNC machining
- Manual milling, turning, and grinding
- High-speed machining
- Precision diamond turning
- Precision welding and soldering
- Electrical discharge machining
- Precision diamond turning

Manufacturing Engineering Laboratory Facilities

The Advanced Manufacturing Systems and Networking Testbed (AMSANT)

This testbed is a distributed, multi-node facility that enables collaborative development and testing of manufacturing systems integration specifications intended to support distributed and virtual manufacturing enterprises.

Manufacturing Software Interoperability Support:

This support is provided through an integration test bed that contains commercial off-the-shelf (COTS) manufacturing and engineering software applications for simulation and visualization.

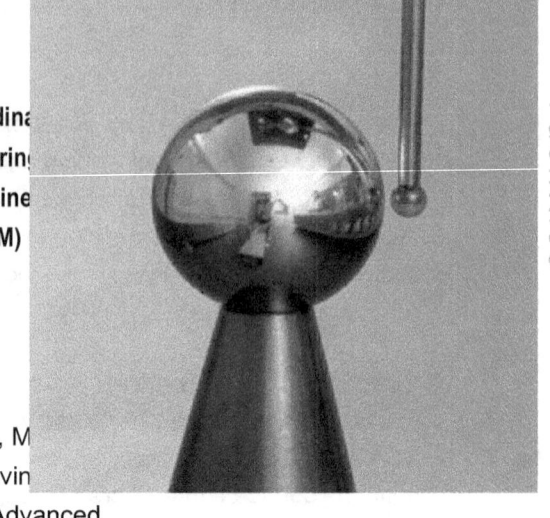

M48 Coordinate Measuring Machine (CMM)

In FY2004, MEL began moving into the new NIST Advanced Measurement Laboratory (AML). The AML features unique, world-leading laboratory facilities in air quality and temperature, vibration, and humidity control that are critical for MEL's precision measurement capabilities. Once the move is completed, 27 out of 41 MEL measurement service areas and 57 major equipment items will be housed in the AML.

The AML offers an unprecedented combination of features designed to virtually eliminate environmental interferences that undermine research at the frontiers of measurement science and technology. The 49,843 square-meter (536,507 square-feet) AML houses 338 reconfigurable laboratory modules and a Class 100 cleanroom - the 8,520 square-meter Nanofabrication Facility. No other facility of this size has so successfully achieved the combined features of strict temperature and humidity control, vibration isolation, air cleanliness, and quality of electric power, some metrics for which are listed on the next page.

Manufacturing Engineering Laboratory Facilities

- Air quality – Class 1000 or better to Class 100 or better
- Temperature – Baseline temperature control within ±0.25 °C to within ±0.1 °C or ±0.01 °C for 48 precision temperature-controlled laboratories
- Vibration – Baseline velocity amplitude of 3 mm/s, down to 0.5 mm/s or less in 27 low-vibration modules – 15 to 100 times better than in NIST's general purpose laboratories
- Humidity – From a baseline control of ±5 % down to ±1 % in special laboratory sections – compared to ±20 % in NIST's general purpose laboratories
- Electrical power – AML-wide uninterruptible power supply prevents outages and counters voltage spikes, drop-outs, and other "dirty power" problems that limit accuracy and precision, reduce analytical sensitivity, and cause long-running experiments to crash.

NIST Advanced Measurement Laboratory (AML)

MEL takes great pride in its facilities and the results that are possible because of them. MEL welcomes collaborators who would like to partner with us and take advantage of these facilities.

MEL Programs at a Glance

Programs at a Glance

Programs-at-a-Glance

In FY2004 the MEL Management Council worked to ensure the future vitality, quality, and productivity of the laboratory's portfolio of technical programs. The MEL Management Council assessed the technical direction and goals, the structure, and the operation of the entire MEL technical program portfolio and made changes as warranted by that assessment. The result is a refreshed portfolio that addresses the needs of our customers, is aligned with NIST strategic directions, makes maximum use of our technical capability, and is managed effectively and efficiently.

This section is designed to give you a quick overview of MEL's eight technical programs. It contains one-page, high-level descriptions for each programs. These descriptions include contact information for the program manager, the program goal, a problem statement, technical approach, and a listing of typical customers and collaborators. A detailed description of each program can be found later in this book and on our website at www.mel.nist.gov/proj.

MEL Programs at a Glance

Dimensional Metrology

Program Manager: Steven D. Phillips

Phone: 301-975-3565

Email: steven.phillips@nist.gov

Program Funding: $4 M

FTEs 19

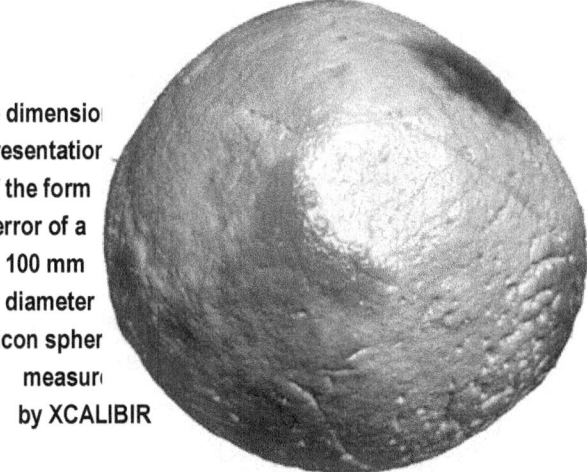

Three dimensional representation of the form error of a 100 mm diameter silicon sphere measured by XCALIBIR

Program Goal

Develop and deliver timely dimensional measurements and standards to address critical U.S. industry needs for traceable dimensional metrology, particularly for the support of trade and innovation, process control, and quality in manufacturing from the micro- to the macro-scale.

Problem

Dimensional metrology spans a vast array of products and industries, from large scale manufacturing of ships and aircraft, to miniaturized mechanical components, to precision optics. The globalization of manufacturing has resulted in complex products with components produced all over the world that must assemble and function seamlessly. Accurate dimensional metrology is essential to meeting this goal. New manufacturing technology is entering industry continuously and often has little or no supporting metrology. NIST, positioned at the top of the traceability pyramid, is challenged to support this broad network of industrial and laboratory dimensional measurements.

Approach

The dimensional metrology program seeks to realize and disseminate the SI (Système International d'unités) unit of length across a wide range of calibration services and artifacts, and to support industrial dimensional metrology through national and international measurement standards. The program funds calibration services and standard reference materials, along with their associated quality assurance programs. The program also identifies and supports research and development (R&D) for new and emerging measurement needs including: (1) precision optics with emphasis on aspheric and free form optics; (2) 3D coordinate measurement of miniaturized features; and, (3) complex 3D surfaces including large scale optical metrology, scanning probe evaluations, and data fitting.

Typical Customers and Collaborators

Department of Defense (DOD), Department of Energy, aerospace industry, automotive industry, heavy equipment and machinery industry, state weights and measures labs, and metrology instrumentation manufacturers.

For more information see page 31

MEL Programs at a Glance

Homeland and Industrial Control Security

Program Manager:	Al Wavering
Phone:	301 975 3461
Email:	albert.wavering@nist.gov
Program Funding:	$3.5 M
FTEs:	8.8

Program Goal

Develop and apply MEL capabilities, tools, and methods to enhance: Preparation for, prevention of, defense against, and response to threats and aggressions against the domestic population and infrastructure of the United States; and Effectiveness of domestic emergency response and law.

Problem

The Department of Homeland Security (DHS) and federal, state, and local emergency response agencies and personnel need an integrated infrastructure of performance metrics, test methods and standards to 1) enable specification, evaluation, and integration of homeland security and public safety equipment and systems, and 2) encourage investment in homeland security science and technology efforts. The lack of consensus in national requirements and standards slows development and leads to confusion among both suppliers and users of homeland security technologies. In coordination with related efforts internal and external to NIST, the MEL Homeland and Industrial Control Security (H&ICS) Program helps build this standards infrastructure, primarily in areas that cut across multiple vulnerability, threat, and response mode categories.

Securing industri[es] including wat[er] electrical power and chemical i[s] a top priority

Approach

In general, pr[ojects in] this program follow the approach used for all DHS standards development and implementation projects. For each component or system under consideration, guidelines will be developed as a collaborative effort among tool developers, users/subject matter experts, and standards experts. Requirements and guidelines will be defined using information related to the capabilities – and the limitations - of thecomponents, and on the conditions in which the component and system are expected to operate. The guidelines will be the foundation for constructing performance measures and testing and evaluation protocols that will provide a reproducible method for assessing and comparing the effectiveness of each system component and of the systems supporting homeland security. Performance measures will encompass: basic functionality, adequacy and appropriateness for the task, interoperability, efficiency, and sustainability.

Typical Customers and Collaborators

Department of Homeland Security Science & Technology and Emergency Preparedness and Response Directorates, National Institute of Justice, ISA: the Society for Instrumentation, Systems, and Automation, Institute of Electrical and Electronics Engineers (IEEE), Critical Infrastructure owner-operators and suppliers, and the Department of Defense.

For more information see page 44

Intelligent Control of Mobility Systems

Program Manager: Maris Juberts
Phone: 301 975 3424
Email: juberts@cme.nist.gov
Program Funding: $4.1 M
FTEs: 13.5

Program Goal

Provide architectures and interface standards, performance test methods and data, and infrastructure technology needed by U.S. manufacturing industry and government agencies in developing and applying intelligent control technology to mobility systems to reduce operational costs, improve performance and safety, and save lives.

Problem

As mobile systems become more intelligent, their use in the field increases. Material handling systems could be used for loading and unloading of trucks, autonomous vehicles can provide an unmanned ground force for the Army, and on-vehicle crash avoidance systems may become more effective. However, to develop and use intelligent mobile systems, industry and government agencies need architectures and interface standards to insure interoperability, real-time sensing technologies for measurement and control, and metrics for evaluating the performance of components and systems

Approach

This program will provide industry with standards, performance metrics, and infrastructure technology to broaden the use of advanced perception and autonomous navigation techniques; provide defen agencies with control system architectures, advanced sensor systems, research services and standards to achieve efficient use of unmanned ground vehicles in the battlefield; provide the evaluation and measurement methods, testing procedures, standard reference data needed to support the deployment of advanced technology in transportation and industrial safety systems. The program will use the NIST-developed reference model architecture for intelligent unmanned ground vehicle - 4D/RCS - architecture as an example of an open system architecture for building complex autonomous robotic systems for other government agency programs. Relevant advanced robotics technology will be transferred to industrial applications.

Typical Customers and Collaborators

- Army Research Lab
- Applanix
- Robotics Technology Inc.
- University of Maryland
- DARPA
- National Highway Traffic Safety Administration (NHTSA)
- The Boeing Company
- Transbotics
- Drexel University
- Visteon & Assistware
- Bremen Univ./Germany
- DOT
- ATR
- TACOM/TARDEC
- EarthData Inc.
- University of Delaware
- OSD/JRP
- General Dynamics Robotic Systems

For more information see page 54

MEL Programs at a Glance

Manufacturing Interoperability

Program Manager: Steven Ray
Phone: 301 975 3508
Email: steve.ray@nist.gov
Program Funding: $3.9 M
FTEs: 26

Program Goal
Equip U.S. manufacturers with the technical guidance and testing support needed to interoperate in today's global, heterogeneous manufacturing world.

Problem
Globalization is the major trend in manufacturing today — globalization of markets and globalization of partners. Both have led to an explosion in the amount of information sharing that must take place. Nevertheless, humans still provide the bulk of the understanding needed to determine what the information means and the majority of the tacit knowledge needed to make decisions based on that understanding. It is absolutely critical to the success of companies and their suppliers that this sharing is done correctly, efficiently, and inexpensively. Changes in technology are positively impacting the way in which this information sharing takes place.

Approach
We work with industrial partners to overcome the barriers that arise from the increased reliance on electronic information exchange, using a virtual manufacturing environment where vendors and manufacturers can test conformance to existing standards; and researchers can validate the next generation of standards incorporating semantic web technologies. This program focuses on three major thrusts: an interoperability testing and demonstration infrastructure; testing of key integration standards for today's manufacturers; and developing semantic technologies for tomorrow's integration needs.

Typical Customers and Collaborators

Industry:
Accordare, Drake Certivo, Lockheed Martin, Nyamekye Research and Consulting Firm, Covisint, General Motors (GM), Ford, Lear, Lesker Corporation, The Boeing Company, Deere & Company, LK Metrology, Mitutoyo, Pratt & Whitney, DaimlerChrysler, GE, LK, Zeis, Nihon Unisys

Consortium:
Automotive Industry Action Group (AIAG), PDES, Inc., Metrology Automation Association (MAA)

Software Vendor:
AutoSimulation, Inc., EDS, Promodel Corporation, Micro Analysis & Design Incorporated, Softimage, Proplanner, Flexsim Software, Emergis, Fujitsu, QAD, SAP, Sterling Commerce, iConnect, Wolverine Software, Simul8 Corp., Delmia Corporation, Rockwell Software, Sewickley, Knowledge Based Systems, Inc., Technomatix, Delmia, Wilcox, Theorem Solutions

For more information see page 66

MEL Programs at a Glance

Manufacturing Metrology and Standards for the Health Care Enterprise

Program Manager: Ram D. Sriram
Phone: 301 975 3507
Email: sriram@nist.gov
Program Funding: $595K
FTEs: 2.4

Program Goal

Apply proven MEL manufacturing technology and expertise to healthcare systems, biomedical devices and equipment, and biomedical data management.

Problem

Spending on healthcare in the United States was about 13.2 % of the Gross Domestic Product (GDP) in 2000 and continues to grow at the rate of 7.3 % per year. Typically U.S. employers offer health insurance benefits to their staff and retirees, making the issue of escalating healthcare costs a major concern. As these costs increase, they raise the cost of doing business and impede our ability to compete globally.

Healthcare and manufacturing share many similar organizational and informational issues. Thus, the healthcare industry as a whole is a customer for the metrology, standard-setting support and technology approaches and solutions that MEL has developed for the manufacturing sector. The healthcare industry

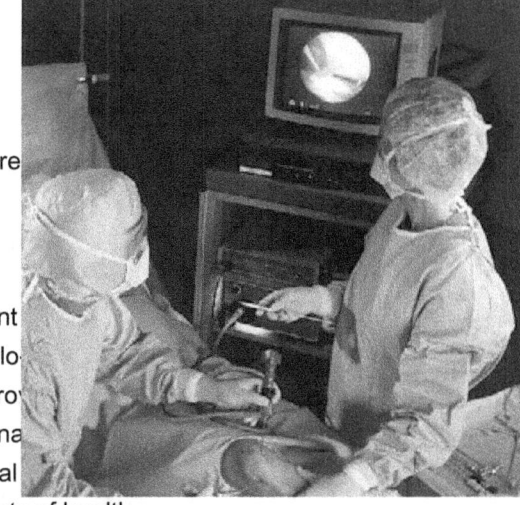

needs an infrastructure that will accelerate and enrich development of methodologies to improve organizational informational for all aspects of health care delivery.

Approach

There are two dimensions to the program: (1) Healthcare informatics; and (2) Medical devices. Healthcare informatics deals with all the processes or "software" of the healthcare enterprise. Medical devices deal with all the products or "hardware" of the enterprise. The program deals with the following objectives within the above two dimensions:

Healthcare informatics
   Enterprise modeling and simulation, Design and production of pharmaceuticals, Biosurveillance, Manufacturing and value chain management

Clinical informatics
   Bioinformatics, Medical devices, Mobility devices, Hearing devices, Intelligent assistive surgical devices (medical robots), Surface characterization of biomedical devices, Meso-micro-biodevices, Nano-biodevices

Typical Customers and Collaborators

Healthcare providers and organizations; Process modeling vendors; Healthcare informatics vendors and consultants; Medical device industry; Academic institutions; Government organizations; Various associations and standard bodies.

For more information see page 79

MEL Programs at a Glance

Mechanical Metrology

Program Manager: Zeina J. Jabbour
Phone: 301 975 4468
Email: zeina.jabbour@nist.gov
Program Funding: $2.3 M
FTEs: 9.3

Program Goal

Develop and deliver timely measurements and standards to address identified critical U.S. industry needs for traceable mechanical metrology in the areas of acoustics, force, mass, and vibration, particularly for the support of trade and innovation, process-control, and quality in manufacturing.

Problem

The Mechanical Metrology Program maintains, realizes, and disseminates the SI units of sound pressure, force, mass, and acceleration to a broad customer base that covers numerous industries and impacts nearly every sector of the U.S. economy. The program staff must maintain and continuously enhance the high-quality, state-of-the-art measurement capabilities, and develop new research areas and measurement services to advance the state-of-the-art in mechanical metrology, provide new opportunities, and boost the competitiveness of the U.S. industry.

U.S. National Prototype Kilogram housed at NIST

Approach

The program will draw upon the internationally recognized skills and expertise of MEL staff in the mechanical metrology areas to develop new and improved measurement capabilities, provide high-quality measurement services, and guarantee open worldwide markets to U.S. industry by participation in international comparisons and standards committees. By increasing the emphasis on Research and Development (R&D), the program will respond to the customer needs for future measurement capabilities and increase the interactions with mechanical metrology R&D organizations and the "end-users" of devices and artifacts calibrated by MEL measurement services.

Typical Customers and Collaborators

Aerospace industry, automotive industry, construction industry, nuclear power industry, pharmaceutical industry, instrument manufacturers, university research labs, state weights and measures labs, federal agencies (Departments of Agriculture, Commerce, Defense, Energy, Labor, Veterans Affairs, and Justice, and the Food and Drug Administration).

For more information see page 86

MEL Programs at a Glance

Nano-manufacturing

Program Manager: Michael T. Postek
Phone number: 301-975-2299
Email: postek@nist.gov
Program Funding: $5.0 M
FTEs: 25

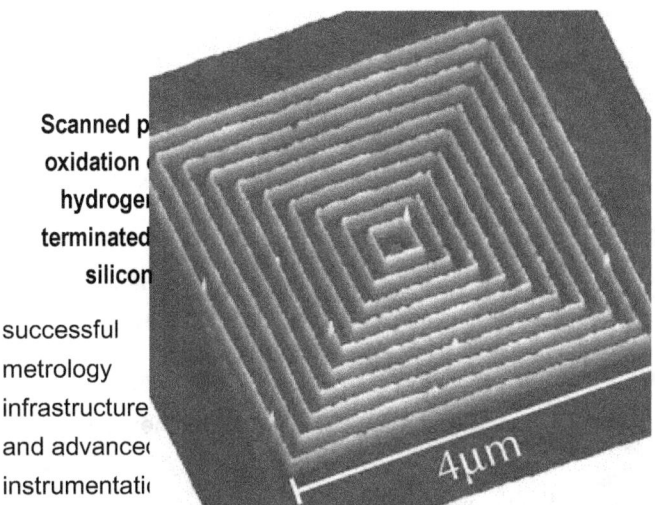

Scanned p
oxidation
hydroge
terminated
silico

Program Goal

Develop and deliver timely measurements, standards, and infrastructural technologies that address identified critical U.S. industry and other government agency needs for innovation and traceable metrology, process-control and quality in manufacturing at the nanoscale.

Problem

Advanced nanomanufacturing is key to the strength and future growth of the U.S. manufacturing sector and a strong measurements and standards infrastructure is vital for its success. It has been predicted that within the next 10 years, at least half of the newly designed advanced materials and manufacturing processes will be built at the nanoscale. Measurement science (metrology) and advanced instrumentation are essential for nanomanufacturing. Metrology is a key enabler for all manufacturing. If you cannot measure it, you cannot make it and that statement is even more accurate in the regime of nanotechnology. Successful metrology infrastructure is essential for manufacturers to achieve the real promise of developing and manufacturing new nanomaterials, devices, and products. Advanced instrumentation provides the necessary data upon which sound scientific conclusions can be based and correct metrology provides the ability to interpret those data properly and accurately. Together, a successful metrology infrastructure and advanced instrumentati tate nanomanufacturing. As pointed out in the National Nanotechnology Initiative (NNI) Instrumentation Grand Challenge workshop final report, some of these metrology techniques will be evolutionary and some will be revolutionary. With that in mind, it is imperative that this program remain agile and evolve with the nanomanufacturing industry and adapt as new applications develop.

Approach

The main theme of this research is "Developing the Nanometrology Infrastructure for Nanomanufacturing." The three fundamental thrust areas are: 1) Imaging and Metrology, 2) Nano-Fabrication, and 3) Control and Assembly Each of these areas addresses unique aspects regarding nanometrology infrastructure. The unique integration of these thrust areas into this program facilitate knowledge exchange to maximize the outcomes of the program.

Typical Customers and Collaborators

- University of Maryland
- Advanced Micro Devices
- Soluris
- International SEMATECH
- Hitachi High Technologies
- FEI Company
- E. Fjeld Company
- NASA
- Spectel Research
- INTEL
- IKLA Tencor
- Semiconductor Research Corp.
- Photronics and Dupont Photomask
- Nova Metrology tools
- Zyvex Corporation

For more information see page 98

MEL Programs at a Glance

Smart Machining Systems

Program Manager: M. Alkan Donmez
Phone number: 301-975-6618
Email: alkan.donmez@nist.gov
Program Funding: $2.8 M
FTEs: 9

Program Goal

Develop metrology methods and standards that enable U.S. industry to characterize, monitor, and improve the accuracy, reliability and productivity of machining operations, leading to the realization of autonomous smart machining systems.

Problem

Coalition on Manufacturing Technology Infrastructure (CTMI) identified an urgent need for "enabling dramatic improvements in the productivity and cost of designing, planning, producing, and delivering high-quality products within short cycle times." CTMI identified thrust areas in process definition and design, smart equipment/process control, fundamental understanding of process and equipment, health monitoring/assurance and integration framework. It also stated that metrology and standards are key enablers of these thrust areas. This program aims to facilitate the development and validation of such measurement methods and standards. A successful program will enable the smart machining systems to cost effective manufacture the first and every part to specification and on schedule.

Modern metrology instruments allow development of new machine tool performance characterization techniques

Approach

The program focuses on developing a methodology for seamlessly integrating all science-based understanding or representation of material removal processes and machining system performance to carry out dynamic and global optimization. There are three programmatic focus areas: (1) performance characterization and representation; (2) process optimization and control; and (3) condition monitoring.

Typical Customers and Collaborators

Boeing, Caterpillar, Pratt& Whitney, Cincinnati Lamb, Hardinge Brothers, Bosch-Rexroth, Ford, Northrop Grumman, Optodyne, Roy-G-Biv, Third Wave Systems, VulcanCraft, American Petroleum Institute, Lion Precision, Heidenhain, Renishaw, Gibbs Associates, IQL, Association for Manufacturing Technology (AMT), The Integrated Manufacturing Technology Initiative (IMTI), U.S. Army, National Nuclear Security Administration (NNSA), National Aeronautics and Space Administration (NASA), University of Massachusetts, University of North Carolina Charlotte, University of Auckland, University of Aachen, University of Maryland.

Programs of the Manufacturing Engineering Laboratory

Programs

Programs of the Manufacturing Engineering Laboratory

MEL Program-Division Cross-Reference Chart

Legend: ○ Primary, ■ Secondary

Programs	MEL						NIST							
	PED	MMD	MSID	ISD	FTD	PL	EEEL	MSEL	ITL	CSTL	BFRL	TS	MEP	ATP
Dimensional Metrology	○	■	■		■							■		
Homeland and Industrial Control Security		■	■	○			■	■	■		■	■		
Intelligent Control of Mobility Systems			○	○					■		■			
Manufacturing Interoperability	■	■	○	■			■		■	■	■			
Manufacturing Metrology and Standards for the Healthcare Enterprise	■	■	○	■					■					
Mechanical Metrology		○			■	■			■			■		
Nanomanufacturing	○	■			■		■	■			■	■		
Smart Machining Systems	■	○	■		■	■	■	■	■					■

Programs

Manufacturing is a complex profession and all U.S. companies, both large and small, are in a furious global competition to survive. U.S. companies face several challenges. Access to global markets depends increasingly on their ability to comply with international standards. Engineers must consider a product's entire lifecycle, that is, taking a concept from design to production to verification and consider the product's eventual disposal. Thousands of different measurements are needed from the time the product is on the drawing board until it is retired from service. No longer does one manufacturer produce every component of a product. Rather, they contract out services and buy components from around the world as needed to manufacture a product. Computers and software applications are now an integral part of the whole process.

A global economy brings challenges, and U.S. companies must seek ways to improve processes, reduce production time, and comply with measurement traceability standards. MEL programs give U.S. companies and manufacturers the tools they need to enhance their competitiveness in the global market. MEL researchers work on measurement, integration, process, and interoperability problems so that manufacturers can spend their time on manufacturing. MEL staff help U.S. manufacturers become more productive and meet the demands for superior quality products by solving tomorrow's manufacturing measurement and standards problems today.

Technology-based innovation remains one of the U.S.'s most important competitive advantages. Today, more than at any other time in history, technological innovation and progress depend on NIST's unique skills and capabilities. For example, nanotechnology is poised to revolutionize industries such as the healthcare, manufacturing, computers, and even telecommunication. Just as they do on the macro-scale, consumers of nanotechnology products must have reliable ways to evaluate, compare, and verify products. MEL researchers work with industry to achieve greater efficiency and productivity with improved measurements and standards, both dimensional and mechanical.

In FY2004, the MEL Management Council and staff conducted a comprehensive strategic planning exercise to ensure the future vitality, quality, and productivity of the laboratory's technical programs. The group assessed the technical direction and goals, the structure, and the operation of the entire MEL technical program portfolio and made substantial changes based on that assessment. This effort resulted in eight strategic programs that are highly focused, strongly connected with industry's needs, well aligned with and supportive of NIST's strategic directions, and funded for success. A transition from an original portfolio of 17 programs was not simply a reorganizing of work, but a true transformation to a well-balanced program portfolio that is intended to result in greater impact and higher efficiency. The new program portfolio is designed to make maximum use of MEL staff's technical capability.

Programs of the Manufacturing Engineering Laboratory

The programs that officially began in FY2005 are:

- Dimensional Metrology
- Homeland and Industrial Control Security
- Intelligent Control of Mobility Systems
- Manufacturing Interoperability
- Manufacturing Metrology and Standards for the Health Care Enterprise
- Mechanical Metrology
- Nanomanufacturing
- Smart Machining Systems

Further information on these programs is given in this section. Program descriptions include program goals, objectives, projects, customer need and intended impact, technical approach, accomplishments, typical customers and collaborators, and both standards activities and measurement services (if appropriate). Under "customer need and intended impact" – you will find answers to questions such as: Who are your customers? What problems do they have? What do they need from NIST? What is MEL's unique contribution to this problem? How will a successful program help U.S. customers develop a competitive advantage? A tactical description of how the program goal and objectives will be met is given in the "technical approach & program objectives" area. A brief discussion of the objectives and how they support the program's ultimate goal is also found here. To provide continuity if appropriate, work from a previous program may appear in the major accomplishments section of the newly designed program. Program descriptions also include a list of FY2005 projects, a brief description of each project, and the relationship between the projects and the program objective(s).

Following each program description, you will find a program timeline. These show programmatic themes, principal activities, deliverables, and impacts for the years leading up to the current year, for the current year, and for future years. Within each timeline, different program themes are differentiated by color and are intended to convey the change in program emphasis over time. Activities, deliverables, and impacts are mapped to the appropriate theme by color. Deliverables are actual outputs while the "impact" section describes beneficial effects of the work. If an activity, deliverable, or impact maps on to multiple themes, this will be shown with a blending of theme colors. Items listed on the timeline for future years are intended activities, deliverables, or impacts of the future work. Detailed information about the items listed (both past and future) are given in the program description that precedes the timeline.

You are encouraged to contact the appropriate Program Manager (contact information is given at the top of each program description) if you would like to find out more detailed information on a particular program or any of its projects. Another source for information is the MEL website http://www.mel.nist.gov/ and then click on "programs."

Dimensional Metrology

Program Goal

Develop and deliver timely dimensional measurements and standards to address identified critical U.S. industry needs for traceable dimensional metrology, particularly for the support of trade and innovation, process control and quality in manufacturing from the micro- to the macro-scale.

Program Manager:
Steven D. Phillips

Total FTEs:
19

Annual Program Funds:
$4 M

Customer Need & Intended Impact

Dimensional metrology spans a vast array of products and industries, from large scale manufacturing of ships and aircraft to miniaturized mechanical components to precision optics. The Dimensional Metrology Program, as part of a national measurement institute, plays a crucial rule in providing metrology infrastructure and calibrates approximately 5,000 master gauges, instruments, and artifacts per year. Although the calibrations performed by the program represent a minuscule fraction of the total number of dimensional measurements performed in the U.S. annually, they typically represent the highest level of measurement accuracy and provide a cost effective way to achieve traceability to the SI (International System of units) units.

Under cost cutting pressures, U.S. industry is frequently unable to maintain in-house metrology expertise and increasingly relies on documentary standards to select and maintain metrology equipment. Accordingly, the program actively participates in the development of national standards and represents the U.S. in the development of international standards. At the national standards level, these documents provide tutorial information and enable cost savings for both suppliers and customers of metrology equipment by providing standardized practices and specifications. At the international standards level, the program supplies a high level of metrological expertise, providing a strong voice to address specific U.S. metrology needs in the ISO standardization process, and ensures that non-tariff trade barriers are not erected.

Industrial manufacturing is continually evolving and creating new measurement challenges for NIST. The Dimensional Metrology Program identifies and supports research and development (R&D) to meet these new measurement needs. Competitive pressures on U.S. manufacturing arising from globalization guide this selection process. As developing economies enter a product field they typically displace domestic suppliers. Domestic manufacturing must continually move up the manufacturing food chain out of reach of commodity-based products. Such an effort requires more sophisticated product designs that frequently involve components that have complex geometries or are miniaturized, or both. In optics, this means aspheric and free form surfaces. In discrete parts, this involves complex surface shape. In these cases conventional measurement methods are either not possible or poorly suited for the task.

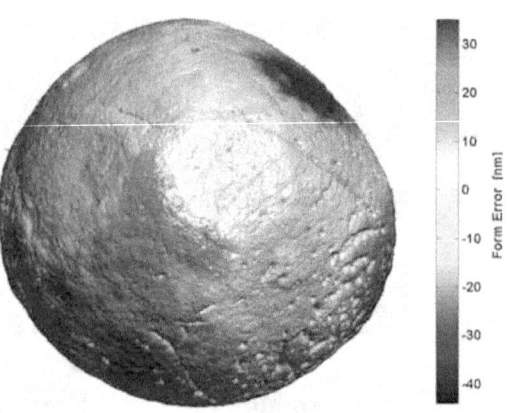

Three dimensional representation the form error of a 100 mm diameter silicon sphere measured by XCALIBIR

Precision Metrology for Complex Optics

Precise optical figure metrology is a leading edge technology, enabling multiple industries. The drive for ever-finer integrated circuit (IC) features demands diffraction-limited imaging; leading-edge steppers seek a wavefront better than 2 nm root mean square (RMS) for adequate imaging. This net wavefront error must be shared between the surface errors of as many as 20 lenses, including aspheric lenses. Wafers on the chuck of the stepper or other tool also need ever-improved flatness to address the depth of field issue. No U.S. SI-traceable metrology system is in place for measuring optical figure, wavefront distortions, or wafer flatness; few industry standards exist, and those that do are often outdated.

Complex surfaces in optics are exploding in their applications. Consumer products such as digital cameras and DVDs are driving the optics industry to use small, fast aspheric optics. Several attendees at the American Society for Precision Engineering (ASPE) meeting on Precision Interferometry in May 2000 identified traceable measurement of aspheric optics as a key need. Traceability involves valid uncertainty statements that are rare in the U.S. optics industry. Only a small number of very specialized optical companies practice the metrology of aspheric and free form optics. The barrier to entry to this field is high in both capital equipment and expertise. The Dimensional Metrology Program seeks to lower this barrier by making this form of optical metrology available to a wide array of smaller manufacturers. The program also seeks to establish a traceability path for optical metrology.

Micro-features

Metrology for micro-features is one of the most active areas in dimensional metrology. The National Measurement Institutes (NMIs) of NPL (UK) and PTB (Germany) are both building custom coordinate measuring machines (CMMs) designed solely for very small parts. Machined micro-features represent a rapidly expanding array of products produced by such techniques as micro-machines and LIGA (a German acronym for Lithography, Electroplating, and Molding). Small features in optical lenses, optical fibers and their connectors, DNA processing chips, drug delivery systems, and a myriad of other applications are increasingly common.

These components are too delicate to inspect with conventional contact probing techniques, including the program's Moore M48 CMM.

New probe technology for microfeatures will provide the measurement infrastructure needed to characterize micro-scale devices entering the marketplace. For example, fuel injectors with holes as small as 60µm in diameter would benefit from precision measurement so as to standardize dimensions, thus increasing fuel economy and reducing pollution; even a 0.4 % increase in fuel efficiency would represent $1 billion annual savings to the U.S. As another example, the $3 billion telecommunication connector market would benefit from the ability to measure the geometry of fiber ferrules or similar fiber alignment devices used in optical switches. Sub-micron alignment is needed to minimize connection losses, and consequently very small measurement uncertainties are desirable. Previously MEL supplied a Standard Reference Material (SRM) for the external diameter of an optical fiber and the program now seeks to provide an SRM for the internal diameter of an optical ferrule. More generally, whereas the $1 billion American micro electro-mechanical systems (MEMS) industry relies on image-based measurements at modest accuracy, we can expect that a maturing MEMS industry will face the same needs for progressive precision as seen in manufactured products on a larger scale; NIST should be in a position to meet this emerging need.

Metallic LIGA component from Axsun Technologies

Complex Surface Metrology

Complex surfaces are increasingly employed in manufacturing, especially for large components. Not surprisingly a boom in instruments and methodologies to measure these structures is underway. A wide range of technologies such as multilateration, photogrammetry, LADAR (Laser Detection & Ranging), and structured light are rapidly advancing due to the availability of high-speed electronics and inexpensive computer power. The U.S. is a major supplier of frameless metrology systems used in scanning large structures, however, demonstrating their metrological capability is problematic. In a 2003 workshop chaired by NPL, PTB, and NIST on large-scale measurement systems, one of the summary findings stated "So far, no common procedures for the evaluation of measurement uncertainty or for performing an interim check are in existence for large-scale measurement systems. In the near future, the rigorous implementation of quality systems, not just in the aircraft and automotive industries, but in a wide context will generate a huge need for action in this area."

The problem is typified by NIST's Building and Construction Research Division (BCRD) that plans to use a three-dimensional (3D) LADAR measurement system to establish the "true value" of a construction measurement test course against which other measurement systems are evaluated. Unfortunately, they have no means to evaluate the accuracy of the LADAR system and, hence, its contribution to calibration errors in the test course. BCRD is very interested in having the LADAR system evaluated in a metrologically rigorous manner.

Complex mechanical surfaces often act as the interface with their environments in dynamic structures such as airframes, turbine blades, and ship hulls. Small deviations in manufacturing or assembly prevent optimal function and cause inefficiencies that can consume large quantities of energy. It is often said that aircraft are actually "powered corkscrews" indicating that small deviations from the designed form create drag that degrades performance. Accurate metrology can minimize these effects. Major manufacturers such as Boeing, Caterpillar, and Pratt & Whitney increasingly rely on measurements of complex surfaces by frameless measurement systems. Traditional methods such as large fixed CMMs represent large fixed capital investments and are not reconfigurable as required in a flexible manufacturing environment.

The program seeks to focus on metrology instruments that are used in advanced manufacturing of complex surfaces. We anticipate providing rigorous calibrations of a wide class of instruments that will facilitate better instrument selection by metrology practitioners, improved designs by instrument manufacturers, and traceability of these measurement systems. The program is also researching conventional CMM probing systems and associated data fitting for complex surfaces.

Technical Approach & Program Objectives

Objective #1: NMI Services

Provide practical access to the SI unit of length and angle through calibrations and standard reference materials, and to support industrial metrology through national and international standards.

The program takes a broad approach to improving all aspects of measurement services and standardization. High quality measurement services are ensured by a program wide quality assurance plan and audits based on ISO 17025. Measurement services include active participation in the BIPM (International Bureau of Weights and Measures, France) sponsored Working Group (WG) for Dimensional Metrology and the international measurement intercomparisons required by the NIST participation in the CIPM (International Committee for Weights and Measures) Mutual Recognition Arrangement. Funding is focused on calibrations with a wide customer base, and less important measurement services are terminated. Program goals include lowering measurement uncertainty, expanding measurement range, and improving on-time delivery. Program staff with specific metrology expertise are involved in both national (ASME - American Society of Mechanical Engineers) and international (ISO TC213 – Technical Committee on Dimensional and geometrical product specifications and verification) metrology standards development. Participation also includes the Joint Committee for Guides to Metrology (JCGM WG1), an international effort that produced the ISO Guide to the Expression of Uncertainty in Measurement. Program staff also provide metrology tutorials to industrial audiences.

Objective #2: Optical Metrology for Complex Surfaces

Develop the instrumentation, methodologies, and expertise to calibrate complex geometry optics - aspherics and free-forms - with sub-nanometer figure uncertainty, for size capacities up to 300 mm diameter for XCALIBIR and 100 mm for GEMM.

Optics is a pervasive enabling technology, and is critical to microelectronics fabrication, consumer electro-optics, remote sensing, defense, medical imaging, and other areas. Worldwide interest in applying aspheric and free form optics is strong, and the need for better metrology in these areas is critical. A key challenge for the measurement of aspheres and free form optics is the necessity to generate a reference wavefront that matches the optical surface under test. The optics used to generate reference wavefronts are called "null lenses." They are costly to make and use, are not reconfigurable, and increase the measurement uncertainty. This objective researches alternative approaches that do not require null lenses: subaperture stitching using both the phase measuring and displacement measuring interferometers of X-Ray Calibration Interferometer (XCALIBIR), the Geometry Measuring Machine (GEMM), and computer generated holograms (CGH). The combination of the three approaches will yield a unique facility for aspheric measurements and provide a means for detecting unknown systematic errors within each method. Additionally, reversal methods that are used to determine the wavefront errors in the interferometer as well as the test optic (and hence are also important to aspheric measurements) are known to have discrepancies between different techniques. XCALIBIR will examine self-calibrating methods and determine their relative uncertainties resulting in improved optical figure accuracy.

Objective #3: Micro-feature Dimensional Metrology

Develop and characterize advanced CMM probing technologies to bring micro-feature (size range: 100 µm – 500 µm) measurement capacity to a level suitable for NIST SP-250 (a publication that lists calibration services) calibrations.

New fabrication technologies are rapidly expanding the array of products with microfeatures. This objective seeks to develop the capability to extend 3D coordinate metrology to the 100 µm feature size, with the added complication of feature aspect ratios of over 50:1. First, several new probe technologies that may be suitable for micro-feature use on the Moore M48 CMM will be investigated. Current work based on a PTB design of a fiber optic probe, in a very rudimentary prototype constructed at NIST, has shown promising results. Alternatively, a MEMS-based probe design (also under development by PTB) will be investigated as a longer-term solution. Different sensor-based technologies will be examined, in particular, for the ability to measure deep holes having aspect ratios approaching 100:1. Secondly, commercial instruments that are pushing this feature size domain, such as Mitutoyo's Ultra UMAP CMM will be investigated under a Cooperative Research And Development Agreement (CRADA) agreement. Thirdly, the program anticipates a CRADA with United Technologies Pratt & Whitney and Oak Ridge Y-12 for the development of measurement methods, reference artifacts, and uncertainty budgets for inspection of small (250 µm to 500 µm) holes in turbine blades. Finally, 2D surface topography will be investigated by both white light interferometry and by phase shifting interferometry microscopy. Some algorithms for stitching interferograms developed under Objective #2 may be used in this effort.

Objective #4: Complex Geometry Dimensional Metrology

Develop metrological capabilities and facilities to calibrate instruments and probing technology capable of collecting large data sets on complex geometry surfaces.

Complex geometry surfaces are among the most challenging areas of dimensional metrology. Some of the difficulties include the complexity of scanning probes, the optical and mechanical metrology in the measuring instruments, huge data files, and fitting algorithms for these surfaces. Most of these instruments are sufficiently transportable (unlike conventional CMMs) that testing at NIST is a reasonable procedure. The program seeks to develop a calibrated test range traceable to the SI unit of length for frameless coordinate measurement systems. Use of the test range by instrument manufacturers may lead to improved accuracy through the discovery of unknown systematic errors.

Scanning probes for conventional CMMs are gaining widespread use for inspecting complex workpieces. These probes are, in effect, miniature three axis CMMs with a very short thermal time constant. All of the thermally induced errors normally associated with CMM structures can also appear in the probe and are accentuated due to the short time constraint. The program will test CMM scanning probes in the NIST Advanced Measurement Laboratory's (AML) thermally controllable lab to simulate conditions on the shop floor.

Finally, the program seeks to develop an E-metrology effort where coordinate metrology information, and in particular, data sets and their associated algorithm fits can be readily accessed by industry. This is an expansion of the current Algorithm Testing System (ATS) into the domain of complex surfaces and large data sets.

Major Accomplishments

Objective 1: NMI Services

- Completed final draft of ASME 89.7.4 "Measurement Uncertainty and Risk Analysis"; this provides the link between measurement uncertainty and the consequences of workpiece acceptance or rejection (FY2004)

- Improved measurement services for ball diameter and roundness, 10 % reduction in uncertainty with no increase in calibration prices (FY2004)

- Publication of ASME B89.7.3.3 "Guidelines for Assessing the Reliability of Dimensional Measurement Uncertainty Statements"; resolves conflicts between supplier and customer concerning the evaluation of measurement uncertainty (FY2003)

Objective 2: Optical Metrology for Complex Surfaces

- Produced and published "Measuring Form and Radius of Spheres with Interferometry" in International Institution for Production Engineering Research (CIRP); one of the first reconciliations of figure measurements with displacement of radius measurements

- Completed design of the Geometrical Measuring Machine (GEMM) world's first 2D interferometer with null free optic for free from optics (FY2004)

- Completed 300 mm diameter 3-flat tests with 0.5 nm repeatability using XCALIBIR, establishing it as an NMI class instrument (FY2003)

Objective 3: Micro-features metrology

- Constructed and tested a 2D optical fiber probe with 15 nm one sigma repeatability; establishes that a 2D probe with $U(k=2) < 0.1$ μm is feasible (FY2004)

- Obtained Mitutoyo Ultra UMAP CMM with micro probe on a CRADA (FY2004)

Objective 4: Complex Geometry Dimensional Metrology

- Completed final draft of ASME 89.4.19 "Laser Tracker Performance Evaluation"; world's first metrology standard for spherically based coordinate measurement systems (FY2004)

- Organized Coordinate Measurement Systems Committee (CMSC) session on Large Scale Metrology Systems; including PTB and NPL coordinating international activities (FY2004)

- Keynote address and paper on large scale metrology to CIRP establishing NIST as a significant contributor to this field (FY2003)

FY2005 Projects

Calibration Services, SRMs, and Intercomparisons (Objective #1)

The Dimensional Metrology Program produces a large number of calibrations and Standard Reference Materials in support of industrial measurement needs. These services are supplied under ISO 17025-based Quality System.

Deliverables for FY2005:

1. A 10 % increase in the percentage of on-time delivery of calibration services.
2. The performance of the Moore M48 CMM for one, two and three dimensional measurements remapped and verified.
3. Calibration service for coefficient of thermal expansion calibration submitted to Measurement Services Advisory Group (MSAG) for approval.
4. Working Group for Dimensional Metrology (WGDM) Drafts for Key Comparison K-6, Diameter.
5. Report of internal audit on MEL's Precision Engineering Division (PED) Quality System. PED Quality System modified to bring SRM development into conformance with the NIST Quality Manual.

National and International Standards (Objective #1)

This project is directed to the development of national and international standards. The list of standards in the section below are standards that program staff either chair or are the primary authors.

Deliverables for FY2005:

1. Publication of ASME B89.7.4 on uncertainty and risk analysis.
2. Final Draft of ASME B89.4.19 for laser trackers
3. Final Draft of ASME B89.7.3.2 regarding measurement uncertainty.
4. Draft International Standard status - ISO 10360-2 on CMMs for measuring linear dimensions.
5. Draft International Standard status - ISO 10360-5 on CMM Probing Systems.
6. Corrigendum to TC 213 regarding U.S. concerns on ISO 14253-1 Decision Rules submitted.
7. Working Draft for ASME B89.4.1 CMM Performance Evaluation compatible with ISO 10360 series.

Programs of the Manufacturing Engineering Laboratory

Optical Metrology of Spheres, Flats, Mild Aspheres (Objective #2)

This project focuses on the characterization of measurement methods and absolute error separation techniques to determine and reduce the uncertainties of measurements made with commercially available Phase Measurement Interferometers.

Deliverables for FY2005:

1. Archival publication demonstrating the capability of XCALIBIR to measure flats and spherical optics with standard uncertainties of 0.2 nm for figure and 20 nm for radius of curvature.

2. Ray-trace model of XCALIBIR to quantify retrace errors that occur during the measurement of aspheres.

3. Figure measurement of near-cylindrical aspheric mirror mandrels for the NASA Constellation-X project.

4. Prepare for move of XCALIBIR to the AML.

Project: Optical Metrology of Aspheres (Objective #2)

This project develops a complementary set of techniques: interferogram stitching, GEMM, and CGH, for the absolute measurement of aspheric optics with uncertainties at the nanometer level.

Deliverables for FY2005:

1. A prototype GEMM instrument useful for 1-D profiling of free form optics.

2. Demonstration of asphere measurement with XCALIBIR stitching algorithms.

3. A set of test aspheres.

Wafer Metrology (Objective #2)

This project will develop an inferred interferometer (IR^2) measuring thickness variation in silicon wafers.

Deliverables for FY2005:

1. IR^2 instrument operational.

2. Demonstration of IR^2 measurement of thickness of 300 mm wafers to less than 5 nm uncertainty. The sample population will include certified 300 mm wafers from several suppliers and single side polished wafers provided by WaveFront Sciences Inc.

Micro-Feature Probe R&D (Objective #3)

The focus of this project is to develop new probe technology for the Moore M48 that can measure feature sizes of 100 µm. The probe requires an uncertainty of $U_{k=2}$ 100 nm and a probing force 20 mN.

Deliverables for FY2005:

1. Paper on 2D fiber optic probe performance published at American Society for Precision Engineering (ASPE).

2. Paper published at National Conference of Standards Laboratories (NCSL) on small hole metrology.

UMAP CMM and Surface Interferometry (Objective #3)

This project focuses on the Mitutoyo UMAP CMM, a state of the art commercial instrument to measure micro-features. This instrument immediately gives the program the ability to measure features as small as 100 μm with $U_{k=2}$ = 300 nm. The instrument will be characterized and serve as a cross reference for ongoing probe development work. Surface interferograms using white light and phase shifting interferometry will also be characterized for surface reconstruction.

Deliverables for FY2005:

1. A paper on the observations and analysis of methods divergence between white light and phase shifting interferometry for surface roughness measurements.

2. A report on the intercomparison of 0.5 mm diameter holes measurements with Moore M48 CMM.

3. Error map and detailed uncertainty budget for micro-feature measurements on the UMAP.

Complex Geometry Metrology Test Range (Objective #4)

This project seeks to develop a calibration test range applicable to a wide range of different measurement technologies including systems that require either optically cooperative or uncooperative targets.

Deliverables for FY2005:

1. Report from a NIST-lead workshop at the CMSC on the calibration needs of large-scale coordinate metrology community.

2. Large Scale Lab refurbished for floor and ceiling work.

3. Report on targets and target mounting.

4. Report on calibration strategies for test range.

High Data Density Probe Performance (Objective #4)

This project seeks to test scanning probes in a simulated industrial thermal environment.

Deliverables for FY05:

1. Integration of UCC1 controller into the REI CMM.

2. Report on performance of integrated probing system when tested with the entire ISO suite of probe performance tests.

E-Metrology of Complex Surfaces (Objective #4)

This project involves construction of the program website, posting as much MEL dimensional metrology information as possible, and the development of downloadable data sets to test metrology software. Data sets will include Algorithm Testing Service (ATS) data, surface metrology data, complex surface data, and Geometric Dimensioning and Tolerancing (GD&T) features with respect to datum reference frames.

Deliverables for FY2005:

1. Surface roughness algorithm database and analysis available on website.
2. Dimensional Metrology Program available for Internal NIST website.

Typical Customers and Collaborators

Customers:
Department of Defense, Department of Energy, aerospace industry, automotive industry, heavy equipment and machinery industry, state weights and measures labs, and metrology instrumentation manufacturers.

Collaborators:
Argonne National Laboratory's Advanced Photon Source; WaveFront Sciences Inc.; Mitutoyo of America; United Technologies Pratt & Whitney; Oak Ridge Y-12; The Boeing Company; Automated Precision Inc.; Renishaw Inc. , and NIST Building Construction and Research Division.

FY2005 Standards Participation

- ASME B89.1.8 Laser Interferometers
- ASME B89.1.9 Gage Blocks
- ASME B89.4.1 U.S. CMM standard including scanning probe technology
- ASME B89.4.19 Optical CMMs including laser trackers
- ASME B89.4.20 Traceability of CMM calibrations
- ASME B89.4.22 Articulating Arm CMMs
- ASME B89.7.3 Measurement Uncertainty: Decision Rules
- ASME B89.7.4 Measurement Uncertainty: Risk Analysis
- ASME B89.7.8 Traceability of Dimensional Measurements
- ASME B46 Surface Texture
- ISO TC213 WG4 International standards on measurement uncertainty
- ISO TC213 WG10 International standard on Coordinate Metrology
- ISO TC213 planetary Dimensional and Geometrical Product Specifications and Verification
- OSA Optical Society of America, "Optical Design and Instrumentation Group"

FY2005 Measurement Services

The Dimensional Metrology Program offers calibration services, including: gauge blocks, tapes and scales, length standards, sieves, algorithm testing, diameter measurements, American Petroleum Institute (API) gauges, optical reference planes, roundness standards, angular measurements, laser measurements, and surface texture.

Programs of the Manufacturing Engineering Laboratory

Dimensional Metrology

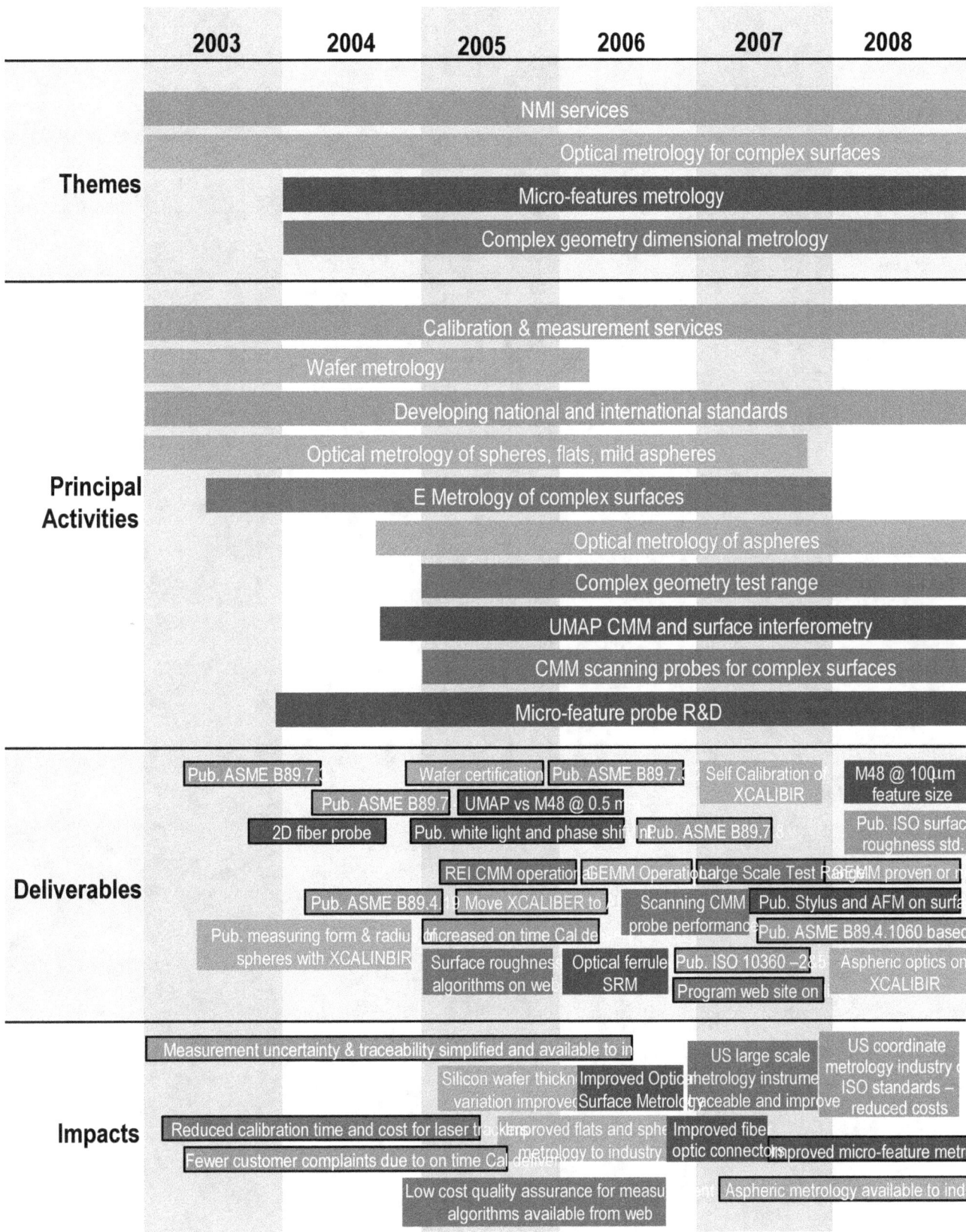

Homeland and Industrial Control Security

Program Goal

Develop and apply MEL capabilities, tools, and methods to enhance:

- Preparation for, prevention of, defense against, and response to threats and aggressions against the domestic population and infrastructure of the United States
- Effectiveness of domestic emergency response and law enforcement

Program Manager:
Al Wavering

Total FTEs:
8.8

Annual Program Funds
$3.5 M

Customer Need and Intended Impact

The Department of Homeland Security (DHS) and federal, state, and local emergency response agencies and personnel need an integrated infrastructure of performance metrics, test methods and standards to 1) enable specification, evaluation, and integration of homeland security and public safety equipment and systems, and 2) encourage investment in homeland security science and technology efforts[1]. The lack of consensus national requirements and standards slows development and leads to confusion in the private sector (e.g., equipment manufacturers, software developers, and shippers) and the government sector (e.g., Federal Emergency Management Agency (FEMA), and state and local emergency planners). In coordination with related efforts internal and external to NIST, the MEL Homeland and Industrial Control Security (H&ICS) Program helps build this standards infrastructure, primarily in areas that cut across multiple vulnerability, threat, and response mode categories.

In addition to federal agency needs, U.S. manufacturers have measurement and standards needs related to securing their own operations and to ensuring that their entire network of supporting services and distributed means of production are able to continue unimpeded. Uninterrupted national and international transportation of parts, subassemblies, and finished goods, continuous flow of power, oil and gas, and chemicals, and robust, secure communication of supply chain information are of paramount importance to manufacturers[2].

---

[1] National Strategy for Homeland Security, Office of Homeland Security, July 2002, http://www.dhs.gov/dhspublic/display?theme=85&content=285.

[2] Manufacturing in America: A Comprehensive Strategy to Address the Challenges to US Manufacturers, US Department of Commerce, Washington, DC, January 2004, available at http://www.manufacturing.gov/.

Programs of the Manufacturing Engineering Laboratory

The ultimate outcome of this program is for federal, state, emergency response, and law enforcement agencies to obtain the best value for their investment in homeland security technologies, to save lives and property through broader deployment of the technologies, and to enable technology developers to accelerate development of advanced capabilities. In addition, many H&ICS projects may result in dual-use benefits. For example, improved knowledge of shipping container contents and location improves security and also improves production scheduling. The H&ICS program responsively addresses new and changing needs of our customers that have arisen post September 11, 2001.

Technical Approach & Program Objectives

In general, projects in this program will follow the same approach used for DHS standards development and implementation projects, illustrated in Figure 1 below. For each component or system under consideration, guidelines will be developed as a collaborative effort among tool developers, users/subject matter experts, and standards experts. Requirements and guidelines will be defined using information related to the capabilities – and the limitations - of the components, and on the conditions in which the component and system are expected to operate. The guidelines will be the foundation for constructing performance measures, and testing and

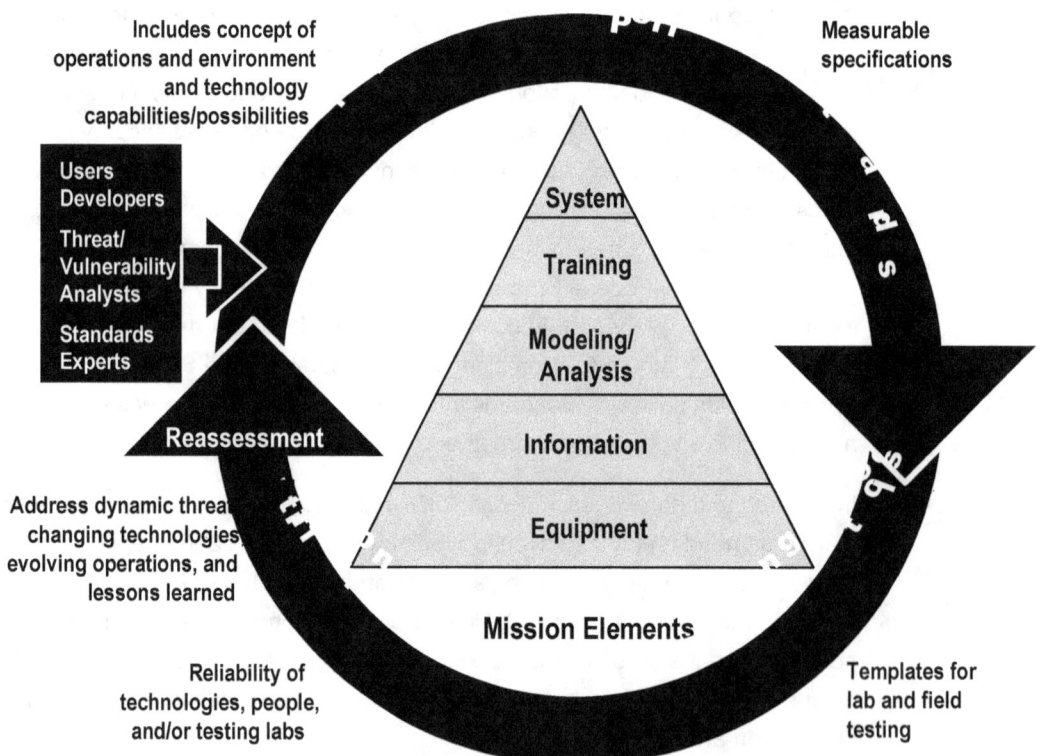

Figure 1. Process for managing DHS standards for all elements of the mission.[3]

---
[3] Source: Bert Coursey, Standards Program Manager, DHS Science and Technology Directorate.

evaluation protocols that will provide a reproducible method for assessing and comparing the effectiveness of each system component and of the systems supporting homeland security. Performance measures will encompass basic functionality, adequacy and appropriateness for the task, interoperability, efficiency, and sustainability.

One of the key parts of the national measurements and standards infrastructure for homeland security hardware, software and processes is the development of consensus performance standards. Standards Development Organizations (SDOs) with relevant interest and expertise are identified for each technical area to be addressed.
This program leverages this existing expertise by establishing a working relationship between these SDOs and other federal agencies to develop consensus requirements and performance standards for system products and processes.

Objectives

Objective #1:
Define and apply performance metrics and standards for homeland security robots,

Objective #2:
Define and apply information security requirements, standards, and test methods for industrial control systems,

Objective #3:
Develop standards for integrating simulation systems and databases for advanced planning, training, and event decision support,

Objective #4:
Lay the groundwork for development of a standards-based framework for homeland security smart sensor networks,

Objective #5:
Demonstrate the benefits of Bullet and Casing Reference Materials for verification of equipment in law enforcement laboratories performing ballistics signature comparisons,

Objective #6:
Evaluate the feasibility of extending the concept of "firearms identification" to develop a national ballistics database for all firearms sold in the U.S.,

Objective #7:
Provide an in-depth understanding of bullet and armor materials leading to predictive models and computer simulations of bullet/armor interactions,

Objective #8:
Complete revisions of standards and guides for emergency vehicle sirens.

Programs of the Manufacturing Engineering Laboratory

Major Accomplishments

- Conducted three urban search and rescue (US&R) Robot Competitions to help accelerate the development of US&R robot technologies: RoboCup2004 U.S. Open Rescue Robot Competition (New Orleans, LA), RoboCup2004 Rescue Robot Competition (Lisbon, Portugal), AAAI2004 Rescue Robot Competition (San Jose, CA)

- Launched DHS-funded multi-year program to develop comprehensive standards and performance metrics for US&R robots

- Released for public comment draft System Protection Profile for Industrial Control Systems, which outlines security requirements to help utilities and other critical infrastructure industries secure their industrial control systems

- Tested compatibility of antivirus software with control system operator interfaces in the Industrial Control System Security Testbed. The test results help industry understand how to select the most appropriate antivirus software configuration settings in an operational control system environment

- Developed Industrial Ethernet performance tests and helped conduct industry-wide testing events, improving the robustness and interoperability of Industrial Ethernet implementations

- Held Modeling and Simulation for Emergency Response Workshop at NIST March 2-3, 2004, resulting in a draft Roadmap for Integrated Modeling & Simulation for Emergency Response

FY2005 Projects

Homeland Security Robot Performance Metrics and Standards (Objective #1)

DHS and other agencies are actively exploring the application of ground, air, and underwater robots to a number of homeland security applications, and need an infrastructure of standard requirements and performance metrics for robot systems and components to help them obtain the greatest benefit from this technology. The requirements will provide concrete performance targets and drive robot development. Performance metrics will give agencies a means to evaluate what they are getting for their money. Without standards and metrics, equipment purchasers have only supplier claims and demonstrations to rely upon. Furthermore, standards are needed to enable robotic technology to be developed on a modular component basis, which enables best-of-breed systems to be assembled. The project has three current thrusts aimed at addressing these needs: 1) developing reference test arenas and conducting competitions for urban search and rescue (US&R) robots (NIST-funded), 2) developing performance metrics and standards for US&R robots (DHS-funded), and 3) defining performance metrics for bomb disposal robots (National Institute of Justice (NIJ)-funded).

Industrial Control System Security (Objective #2)

Foreign adversaries have singled out critical infrastructures as strategic targets for physical and cyber attack.[4] While many security experts agree that physical attacks are the most immediate threat to critical infrastructures, they recognize the need to secure industrial control systems from cyber attack as well.[5,6,7] Two primary categories of standards are needed for industrial control system security. Standard security requirements are needed so owner-operators know what to specify, and so suppliers know what to build. Standard performance test methods are needed to evaluate the impact of security technologies on real-time control system behavior and to measure the performance of control system network components. We are working with other federal agencies and industry representatives in this project to address both of these needs.

A broad cross section of industrial control system users and suppliers participate in this effort through the NIST-led Process Control Security Requirements Forum (PCSRF).

Modeling and Simulation for Emergency Response (Objective #3)

While there are numerous stand-alone modeling and simulation tools for specific homeland security-related domains, these tools need to be brought together for studying the impact of disaster events as a whole. Currently, there are no standards that define the data interfaces, database structures, or software architectures for relevant emergency response simulation and visualization applications. Interoperability standards for simulation and visualization tools for emergency response could significantly improve the Nation's capabilities in this area. An integrated set of simulations could be used for developing well-coordinated response plans. They could be used for providing a complete scenario for training where the results of response actions can be evaluated immediately allowing rapid learning for the trainees. The tools could also be used for rapid evaluation of alternate response plans to a major incident and assist in a prudent selection of the plan leading to minimization of impact from the incident. The needs and a proposed approach to addressing them have been well documented as a result of two workshops on Modeling and Simulation for Emergency Response sponsored by NIST in 2003 and 2004.[8,9] This project builds on the results of these workshops toward establishing a standards framework for emergency response simulations.

---

[4] Unclassified CIA threat briefing at NIST, July 9, 2004.

[5] Challenges and Efforts to Secure Control Systems, GAO 04-354, March 2004

[6] Manufacturing in America: A Comprehensive Strategy to Address the Challenges to US Manufacturers

[7] Making the Nation Safer: The Role of Science and Technology in Countering Terrorism

[8] Modeling and Simulation for Emergency Response: Workshop Report, Standards and Tools, Sanjay Jain and Charles McLean, NISTIR 7071, December 2003.

[9] Roadmap for Integrated Modeling and Simulation for Emergency Response, Sanjay Jain, Charles McLean et al, June 2004 (draft).

Programs of the Manufacturing Engineering Laboratory

Smart Sensor Networks and Technologies (Objective #4)

The development and use of sensors and sensor networks will be critical for the detection of chemical, biological, radiological, nuclear, and explosive weapons and means for their delivery.[10] Currently these systems are being conceived, developed, and deployed on a local/regional basis. Until a common overall standards-based architecture is defined for these systems, it is all but certain that there will be duplication of effort in the engineering and integration required to set them up. Furthermore, if standards are not adopted relatively early in the development and deployment of these systems, they will not be interoperable. This will greatly increase long-term costs and compromise capability and performance. In this new-start project, we will begin to address these issues by identifying user needs and surveying the standards landscape related to smart sensor networks for homeland security. We will identify existing standards that can be applied in the short term, as well as gaps in standards coverage that will need to be addressed in the future. We will also identify where test methods need to be developed for standards validation and conformance. This project will leverage through collaboration existing efforts in this area, including the Oak Ridge National Laboratory's SensorNet program, the Open Geospatial Consortium's Critical Infrastructure Protection Initiative, and Institute of Electrical and Electronics Engineers (IEEE) 1451 (Standard for a Smart Transducer Interface for Sensors and Actuators) working group projects.

Standard Reference Materials for Bullets and Casings (Objective #5)

The National Integrated Ballistics Information Network (NIBIN) is currently under development by Bureau of Alcohol, Tobacco, and Firearms (ATF) and the Federal Bureau of Investigation (FBI). One of the key steps is to establish measurement unification and information sharing in ballistics measurements using the IBIS (Integrated Ballistics Identification System) and other systems. However, to demonstrate completely the reliability and measurement control of these systems, high quality measurement standards for bullets and casings are required. Their key properties include uniformity, reproducibility and measurement traceability. These standard bullets and casings serve as check standards from day to day for each system and will ensure consistency between systems. Now that reference materials for bullets are nearly in place and those for casings will soon follow, the measurement system must be established through laboratory comparisons and refinement of the measurement procedures. In addition, supporting research to improve the cost, chemical stability, and mechanical stability of the standards would be vital to crime labs.

---

[10] Making the Nation Safer: The Role of Science and Technology in Countering Terrorism

Evaluation of Ballistic Imaging Technology (Objective #6)

The purpose of this project is to evaluate the feasibility of the creation and maintenance of a ballistic database, which would consist of digital images and ballistic signature representations of firearms manufactured and imported for sale in the U.S. The technical approach is designed to examine the two fundamental premises of ballistic signature for such a database: 1) the extent to which ballistic markings are unique to individual firearms; and 2) the extent to which the signature of an individual firearm persists over some expected number of repeat firings and remains differentiable from other individuals. The study aims to test the theory of the toolmark signature uniqueness and persistence in the context of topographic measurement of the markings impressed upon surfaces of bullets and shell casings. For that purpose, it will be helpful to use standard bullets and casings being developed by NIST in a separate project to check the reproducibility of different firearm identification and measurement systems. Once the physical basis of toolmark signatures is verified and their dynamics characterized, the study will assess the suitability and limitations of current databases for firearms.

Ballistic Resistant Body Armor and Bullet Studies (Objective #7)

This work involves developing and conducting tests and analytical methods to characterize dynamic properties of bullet and soft body armor materials. The results of these tests will be used to develop a dynamic material properties database for bullet and armor materials, including high rate and heating effects. Special attention will be directed at frangible and other special materials/designs that pose a potential threat to existing armor. Both PBO (Zylon) and Aramid (Kevlar) fiber based soft body armor will be included in the studies. The benefits of this work will be better protective devices for law enforcement personnel, firefighters and other first responders relying on protective equipment. The designers of protective equipment and ammunition will be the initial users of the results of this project. In the future the designers will rely more on modeling and simulation, similar to other manufacturing processes, and accurate material properties will be essential to this modeling effort. Incorporating accurate properties for the materials in the modeling will reduce the number of prototypes that need to be tested and result in a better design more rapidly developed. This in turn will shorten the time from initial concept to equipment in use by law enforcement officers, and dramatically increase the pace at which improved protective equipment becomes available to first responders.

Programs of the Manufacturing Engineering Laboratory

Emergency Vehicle Sirens (Objective #8)

The project will result in improved measurement methods and more comprehensive tests for certifying siren systems and components that are used on vehicles operated by emergency first responders. These sirens provide audible warning that the vehicles are nearby, responding to an emergency, and calling for the right-of-way. These measurements and tests will be developed by government and industry representatives with expertise in acoustical measurements, electrical measurements and/or environmental testing of aftermarket warning equipment for emergency vehicles. Society of Automotive Engineers (SAE) standard SAE J1849 is currently used by organizations that need to cite documents or measurement procedures specified in these documents for certification of equipment used by emergency first responders. Organizations that use emergency vehicle sirens certified to SAE J1849 will have a much higher degree of assurance that these devices have been subjected to a thorough and comprehensive set of rigorous tests and requirements.

Typical Customers and Collaborators

Homeland Security Robot Performance Metrics and Standards

DHS Science and Technology (S&T) Directorate, DHS Emergency Preparedness and Response Directorate/ Federal Emergency Management Agency (FEMA), NIJ, Technical Support Working Group (TSWG), Department of Defense (DOD), National Institute for Urban Search and Rescue (NIUSR), RoboCupRescue League, American Association of Artificial Intelligence (AAAI), Mitre, University of Pittsburgh, University of South Florida, University of Massachusetts, Carnegie Mellon University, National Aeronautics & Space Administration (NASA) Disaster Assistance and Rescue Team, and NIST Information Technology Lab (ITL) and Electrical and Electronics Engineering Lab (EEEL), and Building and fire Research Lab (BFRL)

Industrial Control System Security

The over 500 members in the PCSRF include ABB, Emerson Process Management, Honeywell, Invensys, Rockwell, Cisco, Microsoft, Sun Microsystems, American Gas Association, BP, Chevron Texaco, Exxon Mobil, Association of Metropolitan Water Agencies, American Chemistry Council, Dow, Dupont, Eastman Kodak, Schering-Plough, Georgia-Pacific, General Motors, I-4, National Defense University, Idaho National Engineering & Environmental Lab, Pacific Northwest National Lab, Sandia National Lab, DHS Information Analysis and Infrastructure Protection Directorate and The Society of Instrumentation, Systems, and Automation (ISA)

Programs of the Manufacturing Engineering Laboratory

Homeland and Industrial Control Security

Modeling and Simulation for Emergency Response

Sandia National Labs, University of Southern California, Department of Homeland Security (S&T, Transportation Security Administration, FEMA), National Training Systems Association, Lawrence Livermore National Laboratory, Defense Threat Reduction Agency, Purdue University, Association for Enterprise Integration, National Center for Manufacturing Sciences, Battelle Eastern Science & Technology Center, Defense Modeling & Simulation Office, Argonne National Laboratory, Institute for Defense and Homeland Security, Altarum, University of Southern California, U.S. Air Force, Naval Research Laboratory, U.S. Joint Fighting Command, U.S. Army STRICOM (Simulation, Training & Instrumentation Command), U.S. Army Training - TRADOC (Training & Doctrine Command) Analysis Center, Oak Ridge National Laboratory, U.S. Army RDECOM (Research Development and Engineering Command), and Institute for Defense Analysis

Smart Sensor Networks and Technologies

Oak Ridge National Laboratories, IEEE 1451 member companies, NIST ITL and BFRL

Standard Reference Materials for Bullets and Casings

NIJ, and NIST EEEL/Office of Law Enforcement Standards (OLES)

Evaluation of Ballistic Imaging Technology

NIJ, NIST EEEL/OLES, and NIST Materials Science and Engineering Lab (MSEL)

Ballistic Resistant Body Armor & Bullet Studies

NIJ, NIST EEEL/OLES, and MSEL

Emergency Vehicle Sirens

NIJ, and NIST EEEL/OLES

FY2005 Standards Participation

Homeland Security Robot Performance Metrics and Standards

ASTM International, IEEE, NIUSR, Joint Architecture for Unmanned Systems Working Group, and NIJ

Industrial Control System Security

ISA SP-99 Industrial Control Security committee, American Gas Association AGA-12 committee, International Electrotechnical Commission (IEC) 617842, ISO, ASTM E54 committee on Homeland Security Applications, and Open DeviceNet Vendors Association

Modeling and Simulation for Emergency Response

IEEE High Level Architecture

Smart Sensor Networks and Technologies

IEEE Technical Committee on Sensor Technology/P1451 working groups, and Open Geospatial Consortium

Standard Reference Materials for Bullets and Casings

NIJ

Evaluation of Ballistic Imaging Technology

NIJ

Ballistic Resistant Body Armor and Bullet Studies

NIJ

Emergency Vehicle Sirens

NIJ, SAE

Programs of the Manufacturing Engineering Laboratory

Homeland & Industrial Contr...

	2003	2004	2005
Themes	Performance metrics and standards		
		Standards for public...	
Principal Activities	Homeland security robot p...		
		Industrial contr...	
	Interoperability standards for mod...		
	Ballistics measurements and standards		
		Emergency vehicle siren standards	
Deliverables	US&R robot competitions		
		Sensor network stds roadmap	Standards for US&R and EOD robots
	Industrial control protection profile	SCADA/sector protection profile	
		Control system security performance and interoperability tests	
	SimsER concept prototype		MSER information model
	Characterized standard bullets	Protocol for international comparison of standard bullets & casings	
	Ballistic imaging technology report		
		Ballistic material properties database	
		Revised siren stds	
Impacts	Verification and accelerated development of advanced technologi...		
		Ability to obtain best value for...	
		Improved preparatio[n] defense against, and response to attacks	
		Potential savings of lives and property	

Programs of the Manufacturing Engineering Laboratory

Intelligent Control of Mobility Systems

Program Goal

Provide architectures and interface standards, performance test methods and data, and infrastru(e) technology needed by U.S. manufacturing industry and government agencies in developing and a(pplying) intelligent control technology to mobility systems to reduce operational costs, improve performanc(e,) safety, and save lives. This program concentrates its efforts in the following four main areas:

Program Manager:
Maris Juberts

Total FTEs:
13.5

Annual Program Funds
$4.1 M

Industrial Material Handling & Other Industrial Applications
Reduce costs and improve efficiency in industrial material handling by providing to the industrial autonomous guided vehicle (AGV) industry advanced technology and performance tests to support the use of non-contact safety sensors and appropriate control systems architectures and standards to enable the use of advanced navigation techniques based on such non-contact sensors;

Department of Defense (DOD) Unmanned Ground Vehicles
Save lives and improve national defense capabilities by providing agencies of the Department of Defense (DOD) with control systems architectures, advanced sensor systems, research services, and standards to achieve autonomous mobility for unmanned ground vehicles (UGVs);

Performance Measures for Mobile Robots
Improve vehicle safety, transportation system capacity, and accelerate advancement of mobile robots through the development of advanced sensors and intelligent vehicle control systems on manned and unmanned vehicles by providing performance metrics (i.e., objective evaluation and measurement methods, testing procedures, and standard reference data) needed to analyze sensor and control system effectiveness; and

Knowledge Engineering for Intelligent Control of Mobility Systems
Improve interoperability, enable knowledge reuse, and improve functionality and traceability of knowledge-based mobility systems through the development, implementation, and dissemination of rigorous knowledge capture methodologies, standardized data structures and common knowledge bases.

Customer Need & Intended Impact

To develop and use intelligent mobile systems, industry and government agencies need architectures and interface standards to ensure interoperability, real-time sensing and measurement for control systems, and metrics for evaluating performance of components and systems.

Industrial Material Handling

Discrete part manufacturing customers require material handling systems that can be changed to meet any handling and movement requirements without the need for redesign and manual rearrangement. They also would like a material handling control system that can seamlessly integrate with multiple complex handling systems and one that would self-adapt to changes in handling system configuration or processing requirements. Advanced mobile robot technology is needed in manufacturing enterprises for plant physical security, hazard detection, inventory control, and cleaning. Other industries such as agriculture, construction, postal service and service robots have similar needs. The Integrated Manufacturing Technology Roadmap (IMTR), industry representatives at workshops, conferences, and personal contacts identified the customer needs that this program will address.

Accomplishments from our Intelligent Autonomous Vehicle (IAV) project and our advanced work in autonomous vehicles technology for the military, AGV developers look to us for help in developing and demonstrating advanced capabilities for autonomous loading and unloading of trucks for the warehousing industry. AGVs currently do not provide this capability and the warehousing industry sees this as a high-risk area for vendors to address. There are several strategic advantages of automating this task: gaining a competitive advantage (i.e., performance vs. time); supplementing the workforce since the industry expects to lose between 50 % to 60 % of their labor in the next five years due to aging workers; decreasing by as much as 30 % the warehousing and distribution costs currently incurred through manual loading and unloading of trucks. To perform automated material handling tasks reliably and efficiently with robots, a new generation of advanced, high performance, but low cost imaging sensors are needed by the UGV and other industries. We believe that a new generation of scannerless Focal Plane Array (FPA) LADARs (Laser Detection and Ranging or Laser Radar) will be able to provide this capability.

In addition to advanced technology, the UGV industry needs changes in safety standards to include non-contact bumpers and obstacle detection sensors on their vehicles. Without these changes, AGV users and vendors will be unsure who is liable during vehicle collisions. Although steps have been take to include non-contact bumpers in UGV vehicle standards, a need remains for developing standards for the evaluation of safety system object detection and localization capabilities. Standards will provide the vendors with metrics to develop and users to select systems and system components with certain specified capabilities.

Because of over thirty years of experience in development of technology for intelligent systems, MEL is well positioned to work with the material handling industry and other industries to improve performance, competitiveness and standards to help expand markets.

DOD Unmanned Ground Vehicles

Because of the technological progress in the development and demonstration of autonomous mobility for unmanned ground vehicles in the past decade, the DOD decided to proceed with plans for the deployment of intelligent robotic vehicle platforms in the battlefield by 2010. This program is called the Future Combat Systems program and is jointly managed by the U.S. Army and Defense Advanced Research Projects Agency (DARPA). Standardized architectures and interfaces encourage the use of commercially available "plug-and-play" components and provide reusability and interoperability on a variety of robotic vehicle platforms.

Autonomous tactical behaviors for UGVs are of tremendous interest to the military. With the successful development and demonstration of autonomous driving in the Army's Demo III program, it is now possible to envision the use of these vehicles in tactical situations. However, to implement autonomous tactical behavior functionality successfully in robots, it requires the development of the design methodology, architectural framework, knowledge representation and sensing requirements that support those behaviors. The Army Research Laboratory (ARL) requested that NIST/MEL focus its research to support the need for tactical behaviors for a multi-vehicle route reconnaissance mission.

It is envisioned that a single manned control vehicle could control many autonomous ground and air vehicles to accomplish reconnaissance missions. Successful development of these technologies would result in removing soldiers from performing hazardous missions, multiply the effectiveness of military personnel, help reduce combat casualties, and reduce annual operating costs.

Performance Measures

Companies developing advanced components and system technologies, and government users of such technologies, need objective metrics to evaluate and specify technology elements, products, and intelligent behaviors of complete systems. Such performance metrics improve the efficiency of the development efforts, provide the basis for an equitable and competitive marketplace, and provide the basis of legal and regulatory decisions. The National Highway Traffic Safety Administration (NHTSA) asked NIST/MEL to develop next generation real-time measurement methods, testing procedures, and roadway calibration systems to evaluate the effectiveness of on-vehicle crash avoidance systems for highways. A NHTSA analysis showed that widespread deployment of integrated advanced driver assistance systems addressing rear-end, road departure and lane change collisions could reduce motor vehicle collisions by 17 %. The results from the performance tests will provide the Department of Transportation (DOT) with naturalistic-driving data that will allow them to determine the effectiveness of these systems in preventing crashes, which could save lives and reduce cost.

The DOD needs to conduct Technology Readiness Level (TRL) experiments to determine maturity of military systems and system components in preparation for deployment. In response to Congressional guidance, the Joint Robotics Program (JRP) at the Office of the Under Secretary of Defense (OUSD) initiated plans for developing and fielding a family of mobile ground robot systems for insertion into the military force structure. The JRP organized a collaborative effort to establish a National Unmanned Systems Experimentation Environment (NUSE2) to provide developers and acquirers of unmanned systems with dedicated experimentation facilities, ranges and airspace that is easily accessible. The JRP believes that NIST/MEL has the capability and experience to provide the metrology service for this initiative. If this work is successful, the NUSE2 effort has the potential to provide military service men and women with leap-ahead war-fighting capability, which will also reduce risk levels to military personnel.

Knowledge Engineering for Intelligent Control of Mobility Systems (ICMS)

The development of intelligent ground vehicles (IGV) for the Army objective force requires a thorough understanding of all of the intelligent behavior that needs to be exhibited by the system so that designers can allocate functionality to humans and/or machines. The Army requested NIST to develop an intelligent ground vehicle ontology to capture information about tactical behaviors to enable intelligent behaviors by autonomous systems. The existence of such an ontology is essential to accomplish the level of automation that the community is demanding, which in turn will serve to offload dangerous, human-performed tasks to autonomous systems, thus removing them from harms way. In addition, a fundamental need of mobility systems, in performance of intelligent behaviors, is the ability to have a thorough understanding of the environment. Knowledge representation techniques are needed to capture information that the sensor system perceives, organize that knowledge in a fashion that makes it easy to retrieve, and process.

Technical Approach & Program Objectives

Robotics industry leaders point out that advances in military, transportation, medical, and other non-manufacturing robotics applications, where research and development investments are justified by dramatic potential benefits, will provide the technologies to advance future generation robots for applications in manufacturing. Industrial robots trail in technology development, thus adopting advanced technology developed under military projects is proven to be reliable and cost effective; autonomous mobile systems for military applications represent the forefront of robotics research.

Thus, to achieve the program objectives, the primary technical approach is to use the NIST Real-time Control System (RCS) architecture as an example of an open system architecture for building complex autonomous robotic systems for other government agency programs (i.e., military and transportation), and then transfer relevant advanced robotic technology to industrial applications. The RCS architecture provides a systematic analysis, design, hierarchical framework, and implementation methodology for developing real-time sensor-based control systems. The control system uses sensory information to guide the intelligent vehicle in the execution of complex tasks. Planning for task execution, coordinated activities between vehicles, and for adaptation to changes in the environment are also parts of the total hierarchy.

As intelligent vehicle technology matures and our other agency customers consider deploying intelligent robots for their program tasks and missions, they want to evaluate the performance and effectiveness of this technology as it applies to their requirements and expectations. Based on our history, NIST is in a unique position to fulfill this need for them. However since this is a new emerging metrology field, NIST/MEL will need to put more emphasis on establishing technical strength in performance measures, reference/standards and test facilities that support evaluating the performance of intelligent vehicle systems. Many of the potential sponsors and customers view that measurements and standards are part of NIST's mission. Therefore, some cost sharing for metrology sensors and equipment may increase the likelihood of attracting other agency funds and in establishing long term collaborative relationships. The reference/standard test facilities and measurement equipment are expected to remain at NIST, but could be duplicated at customer experimentation sites.

Programs of the Manufacturing Engineering Laboratory

Objective #1: Industrial Material Handling

Provide industries with the necessary standards, performance metrics, and infrastructure technology to support the use of non-contact safety sensors and control system architectures that enable broader use of advanced perception and autonomous navigation techniques in the AGV, and other industries.

Objective #2: DOD Unmanned Ground Vehicles

Provide DOD agencies with the control system architectures, advanced sensor systems, research services, and standards to achieve next generation autonomous mobility and tactical behaviors for unmanned ground vehicles.

Objective #3: Performance Measures for Mobile Robots

Provide the evaluation and measurement methods, testing procedures and standard reference data for performance analysis and deployment of advanced sensors, and intelligent vehicle control systems on manned and unmanned vehicles used in next generation transportation safety/driver assist systems and in UGVs for the military.

Objective #4: Knowledge Engineering for Intelligent Control of Mobility Systems

Provide rigorous knowledge capture methodologies, standard data structures and common knowledge bases to mobility system researchers and developers to enable improved interoperability, greater knowledge reuse, and increased functionality and traceability of knowledge-based mobility systems.

Major Accomplishments

Advanced Sensing and Control Applied for Industrial AGVs

- We work with a Cooperative Research and Development Agreement (CRADA) partner to develop and demonstrate an autonomous AGV, which uses a LADAR sensor, to locate pallets, pick them up and deliver them into a truck. NIST tested several commercially available LADARs for this application and generated algorithms for detection and localization of pallets.

- We evaluated the performance of prototype next generation FPA LADAR. Developers of these systems project a factor of 10 drop in cost for a LADAR.

- Together with the NIST Building and Fire Research Laboratory, we published a NIST Internal Report (NISTIR 7117) titled "Performance Analysis of Next-Generation LADAR for Manufacturing, Construction and Mobility.". The construction industry recognized the value of this report by announcing its release in a SparView article and in a FIATECH (non-profit consortium focused on development and deployment of technologies in the construction industry) newsletter.

Unmanned Ground Vehicles

NIST Completes Technology Readiness Level Assessment of UGV Performance for the Army Research Laboratory

In the spring of 2003, personnel from the NIST MEL, BFRL and the Information Technology Laboratory (ITL) completed a Technology Readiness Level (TRL) 6 assessment for the ARL autonomous UGV program. The effort used Experimental Unmanned Vehicles (XUVs) developed by General Dynamics Robotic Systems (GDRS). Three similar assessment exercises were conducted in arid, rolling/vegetated, and urban environments. The test vehicles successfully completed over 500 km of autonomous driving. The results show the DOD that the UGV technology developed under the ARL Demo III program is sufficiently mature and reliable to be considered for a variety of tactical missions. The Future Combat Systems (FCS) program considered this when they decided to proceed with the development of an Autonomous Navigation Systems for robotic vehicles.

Adjusted Research and Upgraded Test Vehicles and Mobility Lab Facilities to Support the Next Phase of ARL and Other Agency Mobility Work

Autonomous tactical behaviors for UGVs are of tremendous interest to the military. With the success of the autonomous mobility technology, which was evaluated in the TRL 6 experiments, it is now possible to consider the use of robotic vehicles in tactical situations. Therefore, adjustments and improvements have been made by this program to better support this very demanding new initiative.

- Emphasis has been placed on conducting research to implement autonomous tactical behavior functionalities successfully into robots
- The NIST HMMWV (High Mobility Multipurpose Wheeled Vehicle) testbed vehicle has been upgraded to support research, testing and evaluation of advanced sensors, world modeling, planning and control technology that is needed to demonstrate autonomous performance in tactical behaviors.
- The Mobility Lab has been expanded to accommodate test vehicles for the new DOD and DOT (Intelligent Transportation) research and testing requirements. The lab now accommodates the NIST HMMWV, the GDRS Experimental UGV, and the DOT/NHTSA instrumented test vehicle.

Autonomy Levels for Unmanned Systems (ALFUS) Terminology Document Published

A document defining basic terminology pertaining to autonomous unmanned systems was published by NIST: "Federal Agencies Ad Hoc Autonomy Levels for Unmanned Systems Working Group, Terminology for Specifying the Autonomy Levels for Unmanned Systems, Version 1.0" External groups and committees, including the American Institute of Aeronautics and Astronautics, NASA, NATO and the Canadian Forces requested copies of the report. The seventh autonomy level workshop took place in mid-October 2004.

Performance Measures for Mobile Robots

Road Departure Crash Warning System (RDCWS) Metrics Project Drives On

The DOT/NHTSA funded NIST to provide continued independent measurement and data collection support for the RDCWS Field Operational Tests and to provide planning support for a new four-year initiative in Integrated Vehicle-Based Safety System (IVBSS) program. The IVBSS is a new program to develop and evaluate a multi-function warning system (i.e., road departure, rear-end and lane-change) on multiple vehicle classes (i.e., cars, trucks, and buses).

FY2005 Projects

Industrial Autonomous Vehicles (IAV) Technology Transfer (Objective #1)

This project team will work with key industry AGV developers and user partners to advance the state of autonomous robotic technology.

Deliverable for FY2005:
Demonstration of the autonomous truck loading and unloading with NIST CRADA partner, applying advanced range imaging sensors technology and perception algorithms transferred from DOD and DOT projects.

Deliverable for FY2006:
Collection of data, advance processing algorithms, and design implementation toward integration of advanced 3D sensors onto industrial AGVs and other mobile robots.

ARL Tactical Behaviors (Objective #2)

This project team will support the Army Research Laboratory (ARL) program in tactical behaviors.

Deliverable for FY2005:
A demonstration of next generation perception, planning and autonomous mobility that supports multiple vehicle tactical behaviors using real/virtual simulation environments.

Deliverable for early FY2006:
UGV intelligent tactical behaviors for route reconnaissance available to ARL for Field Operational tests.

NIST Mobility Testbed Vehicle (Objective #2)

This project team will provide DOD agencies with the control system architectures, advanced sensor systems, research services, and standards to achieve next generation autonomous mobility and tactical behaviors for unmanned ground vehicles.

Deliverable for FY2005:
The NIST mobility testbed vehicle upgraded to evaluate and refine next generation perception, learning, planning, and multi-vehicle behaviors for a reconnaissance mission.

Deliverable for FY2006:
A demonstration of next generation autonomous technology and tactical behaviors at NIST.

Military UGV & US Standards (Objective #2)

This project team will support relevant military standards activities.

Deliverable for FY2005 and beyond:
Working with standards groups on standards for the deployment of UGVs in the battlefield.

NHTSA Integrated Vehicle Safety Systems (Objective #3)

The project team will support the NHTSA Intelligent Transportation Systems (ITS) program initiative in Integrated Vehicle Based Safety Systems.

Deliverable for FY2005:
A NHTSA new program plan and Field Operational Tests in Integrated Vehicle Based Safety Systems developed to prevent highway crashes.

Deliverable for FY2008:
Development of the Next Generation baseline measurement system and test method concepts for measuring performance of Integrated Vehicle Based Safety Systems in preventing highway crashes.

Reference/Standards and Metrics for NUSE2 (Objective #3)

The project team will support the OSD Joint Robotics Program office by developing references and standards, and metrics for evaluating the performance of intelligent systems.

Deliverable for FY2006:
Assuming funding becomes available–
An initial baseline reference test facility for small autonomous robots at NIST.

Deliverable for FY2007:
Assuming funding becomes available–
Transition of reference/standard assessment technology to the NUSE2 experimentation sites

Programs of the Manufacturing Engineering Laboratory

DOD/DOT Intelligent Vehicle Technology Transfer (Objective #3)

This project team will conduct a workshop and initiate a process for intelligent vehicle technology transfer between the DOD Joint Robotics Program, the DOT Intelligent Transportation Systems program, and their associated industry and research partners.

Deliverable for FY2005:
Assuming funding becomes available - a workshop and a final report to DOD and DOT describing the results of the workshop.

TARDEC Intelligent Ground Vehicle Ontology Program (Objective #4)

This project team will support the U.S. Army's Tank and Automotive Research, Development and Engineering Center (TARDEC) office in the development of a neutral Intelligent Ground Vehicle Ontology for representing tactical behaviors for unmanned ground vehicles.

Deliverable for FY2005:
An initial version of the Intelligent Ground Vehicle Ontology that focuses on scenarios identified by the TARDEC STOs (Science and Technology Objectives).

Knowledge-based Representations for Mobility Systems (Objective #4)

This project team will develop data structures and knowledge bases that will allow an autonomous mobility system to better understand itself and the environment.

Deliverables for FY2005:
1. Representations for moving objects that include the ability to predict their future location probabilistically.
2. Exploration of tools to aid in knowledge entering, reasoning, and visualization.

Typical Customers and Collaborators

Government:
- Army Research Lab: Tactical Behaviors for Unmanned Vehicles
- DARPA: Future Combat Systems Technology Development
- DOT/NHSA: Metrics for Vehicle-based Safety Systems
- TACOM/TARDEC: Intelligent Vehicle Ontology for Tactical Behaviors
- OSD/JRP: National Experimentation Sites for Unmanned Robotic Systems

Industry:
- Applanix: Integrated (INS/GPS) Position and Orientation Systems
- ATR: Advanced Computing and Simulation
- Transbotics: Automated Loading and Unloading of Trucks
- Visteon & Assistware: Road Departure Crash Warning Systems
- Robotic Technology Inc.: Military and Transportation Robotics
- Advanced Scientific Concepts, Coherent Technologies: Next Generation LADAR
- Lockheed Martin, Raytheon: Next Generation LADAR
- Aeolean: Advanced Computing
- EarthData Inc.: High Resolution Ground Truth
- Automated Material Handling Association: Automated AGVs

Universities:
- University of Delaware: Dynamic Modeling of Vehicles, Path Planning
- Drexel University: Real-time Hierarchical Control Architectures
- University of Maryland: Gesture Recognition, Hyper-spectral Imaging
- Bremen University/Germany: Research in Next Generation Planners
- MIT/Lincoln Labs: Next Generation Solid State LADAR

FY 2005 Standards Participation:
- American Society of Mechanical Engineers (ASME) B56.5 AGV Bumper Standards Committee – participate in committee activities
- ISO Technical Committee 204, Working Group 14 – Standards for Lane Departure Warning Systems – participant
- DOD Joint Architecture for Unmanned Systems – participant
- DOD Weapon System Technical Architecture Working Group – participant
- DOD Vetronic Architecture Standards for Unmanned Ground Vehicles – participant
- Autonomy Levels for Unmanned Systems (ALFUS) – facilitator and participant
- Infrastructure & Standards Committee for Unmanned Systems - participant

Programs of the Manufacturing Engineering Laboratory

Intelligent Control of Mobility Systems

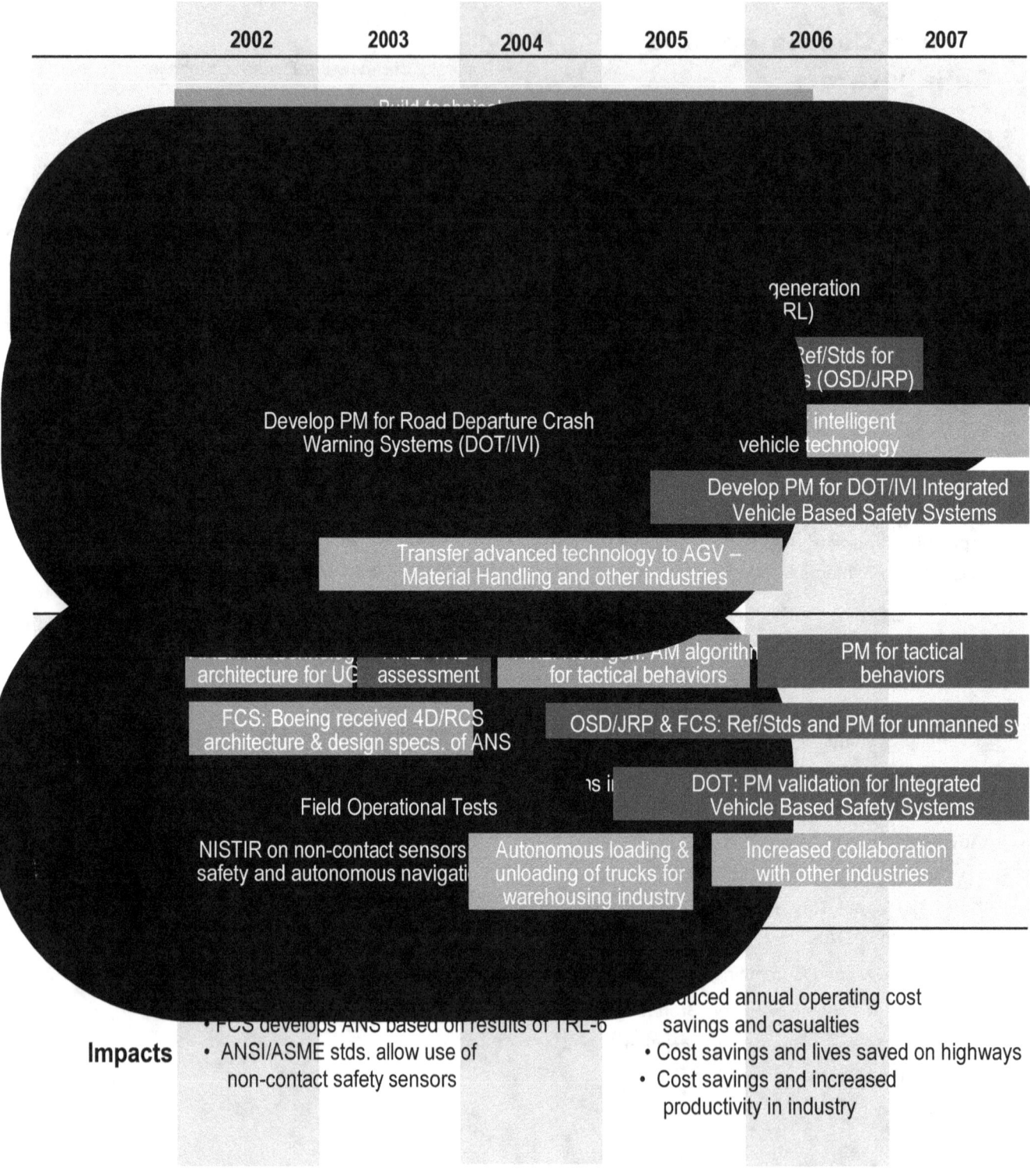

Impacts
- FCS develops ANS based on results of TRL-6
- ANSI/ASME stds. allow use of non-contact safety sensors
- Reduced annual operating cost savings and casualties
- Cost savings and lives saved on highways
- Cost savings and increased productivity in industry

Manufacturing Interoperability

Program Goal:

Equip U.S. manufacturers with the technical guidance and testing support needed to interoperate in today's global, heterogeneous manufacturing world.

Customer Need & Intended Impact

Program Manager:
Steven Ray

Total FTEs:
26

Annual Program Funds:
$3.9 M

Globalization is the major trend in manufacturing today — globalization of markets and globalization of partners. The globalization of markets means that the companies want to sell their products all over the world. The globalization of partners means that supply chain members are also located all over the world. Both have led to an explosion in the amount of information sharing that must take place. It is absolutely critical to the success of companies and their suppliers that this sharing is done correctly, efficiently, and inexpensively.

Changes in technology, from faster networks to new programming languages such as XML (EXtensible Markup Language), are impacting the way in which this information sharing takes place. Nevertheless, humans still provide the bulk of the understanding needed to determine what the information means and the majority of the tacit knowledge needed to make decisions based on that understanding. All of this is about to change with the advent of the Semantic Web. Simply stated, the Semantic Web will enable computers to understand the meaning of concepts, to reason about those concepts, and act on those concepts according to the rules they have been given. The resulting programs will operate at the semantic level, not the data level. They will know that purchase orders are different from schedules, which in turn are different from Numeric Control (NC) programs; and they will know how to deal with those differences.

This program addresses the information sharing needs of Original Equipment Manufacturers (OEMs), their Small and Medium Enterprise (SME) suppliers, and software vendors. In addition, we have customers within the federal government, including the Department of Energy (DOE) Oak Ridge National Labs and the Department of Defense (DOD) departments of the Army, Air Force, and Navy. Within NIST, the evolving Interoperability Test Bed (ITB) is already supporting the work of other laboratories. Those laboratories include the Electronics and Electrical Engineering Laboratory (EEEL), Building and Fire Research Laboratory

Programs of the Manufacturing Engineering Laboratory

(BFRL), Chemical Sciences and Technology Laboratory (CSTL) and Information Technology Laboratory (ITL). Finally, we have a number of customers from the international standards-development sector. We actively participate in technical working groups in three de jure organizations: American Society of Mechanical Engineers (ASME), Institute of Electrical and Electronics Engineers, Inc. (IEEE), and ISO. We also participate in other industry-led standards development organizations: OMG, OASIS, and W^{3}C.

To survive in the global economy, our customers repeatedly emphasize the need to get the right information to the right place in the right form at the right time. Succinctly, their goal is information integration anywhere, anytime. Our customers asked NIST to address three specific and important needs that will enable the realization of this goal.

NIST should:

- Provide methods, tools, and data sets for testing conformance to existing international and de facto standards.

- Evaluate: (1) the standards conformance of key implementations – what was implemented agrees with the specification (conformance testing), (2) that the standards meet the business requirements they were intended to address (validation testing), and (3) that sets of business applications can successfully operate together (interoperability testing).

- Propose a new generation of standards technology that is based on formal semantics, to support both the automation of the integration process and the harmonization of existing, conflicting standards.

NIST is uniquely positioned to respond to these needs. Our service to customers is perhaps best illustrated in Figure 2.

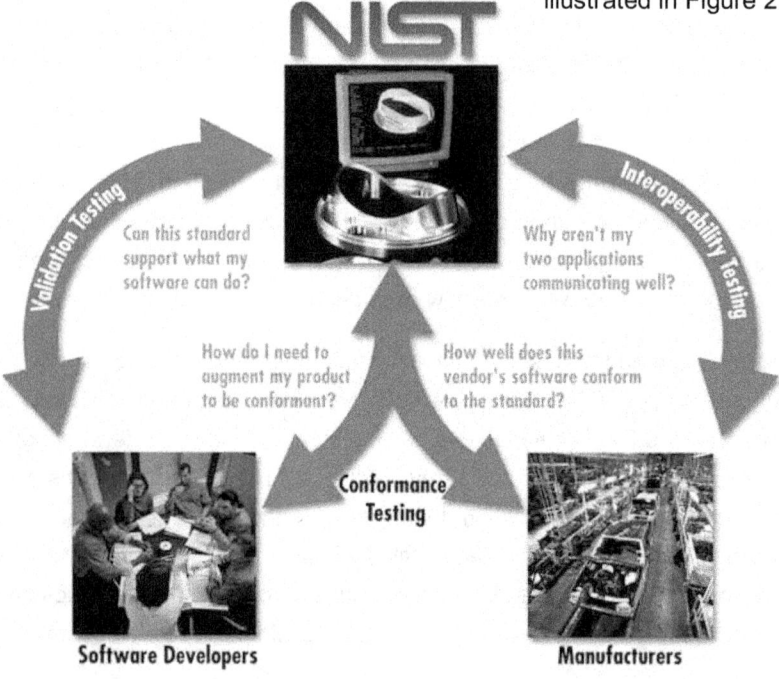

Figure 2. Customer Needs

Technical Approach & Program Objectives

We aim to equip today's manufacturer with the guidance and testing support needed to participate in the global, distributed manufacturing world. We work with industrial partners to overcome the information-handling barriers that have arisen from the increased reliance on electronic information exchange with distant customers and suppliers, using a virtual manufacturing environment where vendors and manufacturers can test conformance to existing standards, and researchers can validate the next generation of standards. A picture of our vision is shown in Figure 3.

Program Thrusts

The Manufacturing Interoperability Program focuses on three major thrusts:

A. An interoperability testing and demonstration infrastructure

B. Testing of key integration standards for today's manufacturers

C. Development of semantic technologies for tomorrow's integration needs

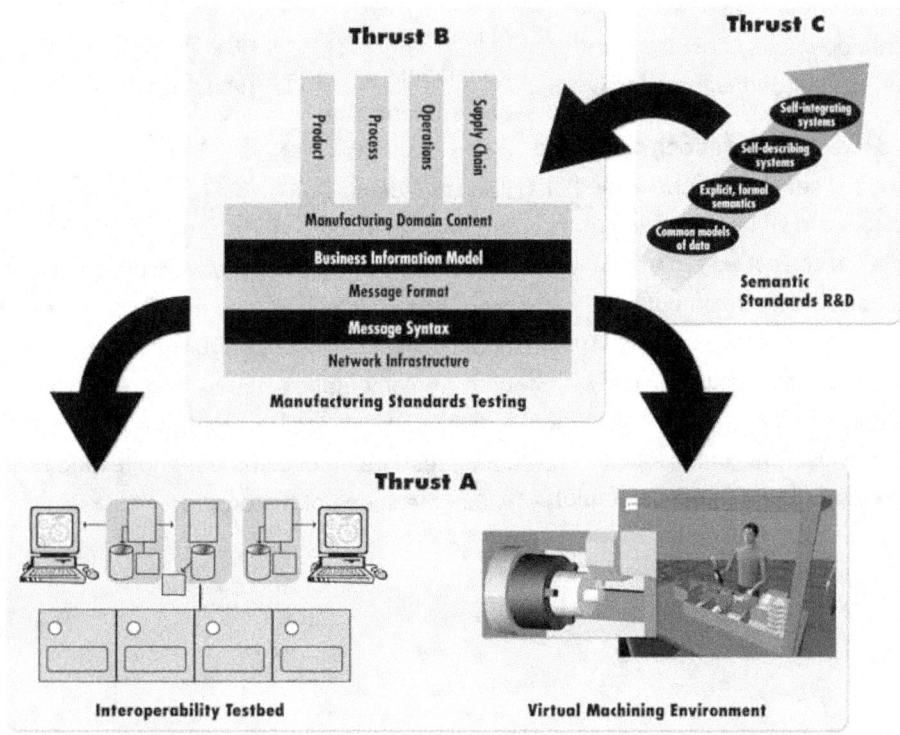

Figure 3. Program Vision

Programs of the Manufacturing Engineering Laboratory

These three thrusts depend on one another to be successful. Integration standards will be identified in concert with industrial partners for key information supporting product, process, operations and supply chains. Pragmatic choices will be made to provide a recommended suite of standards for today's modern manufacturer.

These standards will be supported through the interoperability testing and demonstration infrastructure containing two major components: a testing environment for logging, diagnosis, conformance and interoperability testing focused at the content level, plus a piloting and demonstration environment populated with commercial production software tools to establish the usability of these standards-based approaches in realistic settings. These two components are called the Interoperability Test Bed, and the Virtual Manufacturing Environment, respectively.

Finally, the program is investing in a strong research thrust to support the use of semantic technologies in new standards. It is becoming widely accepted that semantic technology is the correct way to transmit information in an unambiguous and computable fashion, and we are recognized leaders in this arena. The tools within our testing infrastructure will rely upon the semantic research results, and the content-oriented tests will typically use ontologies, either explicitly specified in the standards (most desirable), or reverse-engineered from the standards.

Thrust A - Interoperability Testing and Demonstration Infrastructure

This infrastructure builds upon past efforts and collaborations initiated under earlier MEL programs and projects that addressed interoperability testing and the simulation of manufacturing supply chains, systems, and processes.

Objective #1

Will establish a program Technical Advisory Board (TAB). The Technical Advisory Board will be comprised of leading experts from industry, government, standards organizations, academia, and the research community. The TAB will advise program management on changing industry needs, program directions, collaboration opportunities, evolving technologies, implementation issues, and relevant external activities. It will also document the industry problems, research, development and testing issues within the scope of the program.

> "NIST is doing what our industry members won't do – providing the infrastructure and testing resources to speed the development and implementation of new specifications."
>
> David Connelly, President
> Open Applications Group

Objective #2
Establishes the Interoperability Test Bed (ITB). The Interoperability Test Bed will consist of a testing framework and tools for standards conformance and interoperability demonstration, building on the demonstrated success of the Manufacturing Business- to-Business (B2B) Test Bed and the Metrology Interoperability Test Bed. A report documenting the ITB configuration, tools, methods, and procedures will be produced.

Objective #3
Establishes a Virtual Manufacturing Environment (VME) capability within the NIST laboratories. The Virtual Manufacturing Environment will pilot interoperability solutions, and will be an effective means to communicate the value of this program to decision makers in industry and government. The VME will enable NIST to implement, evaluate, and demonstrate the feasibility of interoperability solutions using real manufacturing software applications, as well as simulations and emulations of manufacturing processes and equipment. VME software systems will be augmented with manufacturing hardware in the MEL Fabrication Technology Division and other laboratories in MEL. Manufacturing hardware may include numerically controlled machine tools, coordinate measuring machines, robots, and other manufacturing equipment. Visualization capabilities within the Advanced Manufacturing Systems And Networking Testbed (AMSANT) Facility will be enhanced to support VME pilot implementations and demonstrations. Web cam and other remote visualization capabilities will be installed to monitor actual physical hardware running at remote locations from the AMSANT. A report documenting the VME configuration, applications, procedures and test data sets will be produced.

Thrust B – Integration Standards Testing

A typical interaction between a manufacturer and a customer or supplier contains a minimal set of information, often called a technical data package. This information includes a specification of a product to be manufactured along with quality specifications and at times processing requirements. The manufacturer in turn must incorporate this manufacturing need into its ongoing operation plans – schedules, inventory, resource and component requirements. By focusing on just this minimal set of requirements, we identified a core set of cross-referencing standards needs to support a majority of manufacturers, especially small manufacturers. These standards – product, process, operations, and supply chain standards as shown in Figure 3 – must be consistent with one another to enable smooth information flow. This program will identify and validate such a consistent standards suite by teaming with industrial partners in the various specializations in an environment of close and frequent communication among the specializations. In this manner we will avoid a disjoint set of standards that cannot support interoperation among them. This is achievable only by virtue of having an interoperability program with the breadth of scope present here at NIST.

The specifications, typical software applications, and associated test data will be validated using testing tools within the Interoperability Test Bed. One or more real world test case data sets will be established for each specification in cooperation with industrial and research collaborators. Test Bed tools will support information modeling, wrapper development, system prototyping, verification, validation, conformance and interoperability testing.

The Virtual Manufacturing Environment (VME) will be used to test-drive the specifications and test data sets with both real and simulated manufacturing software applications. Existing commercial software applications may be used "as-is" or extended to support required interfaces and functionality. If appropriate commercial or research systems are not available, prototype applications may be developed to fill holes in the integration scenario. Due to resource limitations, prototype development will usually only be undertaken as a last resort when it is critical to the effective achievement of an interoperability objective.

Objective #4

Addresses the collaborative identification of a suite of complementary standards supporting the exchange of product, process, operations and supply chain information, with associated testing support. A consistent approach will be used to achieve each interoperability scenario. First, a system architecture and scenario document will be developed. This document will describe industry's requirements, the software applications or functional modules that need to be integrated, and an operational scenario that illustrates how the integrated applications and modules should ultimately function together.

The next steps will be to identify and test interface specifications that will be used to implement the architecture. Working with industry and other research collaborators, test data sets will be assembled that reflect real industry problems. The test data sets will be used to populate software applications, and where appropriate, associated databases. Test data will be used to exercise interfaces, perform various integration tests using tools of the Interoperability Test Bed, and pilot solutions within the Virtual Manufacturing Environment. Demonstrations will be scheduled periodically to show progress towards and completion of integration.

Programs of the Manufacturing Engineering Laboratory

As individual specifications are established or identified, attention will be concentrated on integrating them into larger networks of interconnected systems and interfaces, specifically through the following integrations:

1. Expanded supply chain interoperability through the integration of supply chain business processes specifications with the production management operations specifications
2. Expanded internal engineering data flow through the integration of the engineering product and manufacturing process specifications
3. Full integration across a virtual enterprise through the joining of the combined supply chain specifications with the combined product and process specifications.

The first integration will give supply chain managers and participating suppliers access to both inventory levels and production schedules throughout a supply chain, expanding upon the goals of the Automotive Industry Action Group (AIAG) Inventory Visibility and Interoperability program.

The second integration will provide a consistent, standards-based treatment of the engineering data from concept to manufacturing process execution.

The third integration will provide a consistent set of standards covering the supply chain management, internal production management, production process functions and engineering data flows addressed in earlier work.

Thrust C – Development of New Technologies for Interoperability Standards and Integration

Tools and standards for information system development have gone through a long process of abstraction and layering. Each layer enables individuals to direct these devices in ways closer to how they understand the problem being solved. Early computer systems were programmed by physically wiring computer circuit boards. Machine programming and assembler language raised this to the symbolic level. Later languages such as Fortran and C brought languages closer to simple mathematics, and LISP and Prolog enabled programmers to build intelligent systems that reflected human problem solving techniques.

Despite these improvements, in many respects software standards and technology have changed very little over the last 20 years. The languages have become a bit easier to use, but not vastly different from Fortran of the 1950's. Programmers and architects still sift through libraries of reusable elements, and put them together in a handcrafted way. Many tools are available now, but just as aids in the manual process of software construction. They are comparable to the improvement of a circular saw over a handsaw, rather than robotic and NC manufacturing systems over manually controlled machine tools.

> **"Semantics-based integration tools are destined to become increasingly powerful and capable."**
> From CIO Magazine, August 2002

Programs of the Manufacturing Engineering Laboratory

This thrust will complete the fundamental research necessary for information system development to move to a new level of automation and facilitate the integration of industry's next generation of systems. The components will be:

- Identification of semantic approaches
- Ontology development
- New tools and techniques
- Prototype systems

Semantic approaches are the next logical step in bringing computer languages closer to the way business, engineering and manufacturing experts understand their problems. These approaches express the expert's concepts in terms of meaningful, computable statements. The techniques are not oriented toward computational speed, but rather precision of expression. In this way, the specific conclusions or effects that are expected of the system can be recorded, validated with automated reasoning and simulation, and used to verify the performance of the system eventually built. Its semantic description can be searched by others looking for existing functionality to reuse, or can be automatically composed with semantic descriptions of other systems to create newly integrated ones. The semantic description can also be used to present the data present in the system in a way domain experts can understand, and enable automated and semi-automated decision-making.

> **Ontology capabilities will become a core technology. [...] By 2005, lightweight ontologies (taxonomies) will be part of 75 percent of application integration projects [...] By 2010, ontologies using strong knowledge representations will be the basis for 80 percent of application integration projects.**
> From Gartner, "Semantic Web Technologies Take Middleware to the Next Level," August 2002

Objective #5

To identify the semantic approaches that are appropriate to manufacturing. The primary concern is that these be understandable by experts at the same time they are rigorous enough to enable computer support. In particular, they must be self-documenting and address concrete instances of concepts, so it does not require the original author or extensive documentation for others to understand the meaning of the information, and so they are amenable to commonly available automatic reasoners. It is also important that the approaches leverage existing or near-term information standards to ensure relevance.

Objective #6

To apply the identified semantic approaches to the primary areas of manufacturing: product, process, operations, and supply chain. These will require a common set of primitive concepts that are reusable as needed in the various areas. The results will be amenable to automatic reasoning, to enable application in complex and diverse systems, coordination of knowledge across organizations and disciplines, enhanced support for enterprise design, integration, and decision-making, unambiguous of interoperability standards, and autonomous agents. MEL's experience in both manufacturing and ontologies will apply to these areas.

Objective #7

To develop tools and techniques for working with the ontologies developed in the previous objective. It will address automation in the capture of the knowledge implicit in interfaces to manufacturing software, the design of inter-system processes using those interfaces, and generation of translation software from these specifications. This will involve the matching of concepts between existing systems and requirements on the integrated process, as well as the expression of those concepts in specific implementations.

Objective #8

Apply and test the tools developed in Objective #3, using actual industry cases drawn from the many incompatible standards currently emerging. Part of the testing will be to ensure that the results of the tools conform to existing tests provided for the standards involved. This objective will also identify those aspects where tool support requires further research in the underlying semantic representations.

Major Accomplishments

Metrology Test Bed Launched

The "test bed" is a group of people and physical resources located at NIST in the Manufacturing Engineering Laboratory that support development and testing of standards for components of dimensional metrology systems. Our primary tie to industry is through the Automotive Industry Action Group Metrology Interoperability Team (AIAG-MIPT). For more information, please visit the test bed website:
http://www.isd.mel.nist.gov/projects/metrology_interoperability/NISTactivities.htm#testbed

Simulation Standards Consortium Established

The Manufacturing Engineering Laboratory hosted the Simulation Standards Consortium kick-off meeting at NIST on February 25, 2003. Charles McLean made a presentation on the mission, goals and objectives, and motivations for joining the consortium; and offered a framework of modeling and simulation standards for focusing the efforts of the Consortium. Swee Leong presented the operational plans for this Consortium. A draft Simulation Standards Consortium agreement was distributed to the participants for their review, comments, and approval. Meeting attendance represented 19 organizations from major software vendors, industrial companies, government agencies, and academia. Responses to the Simulation Standards Consortium have been very positive.

Programs of the Manufacturing Engineering Laboratory

"Economic Impact of Inadequate Infrastructure for Supply Chain Integration" Study Completed

This economic impact study examined the current state of supply chain integration, estimated the economic impact of inadequate integration, and identified opportunities for governmental organizations to provide critical standards infrastructures that will improve the efficiency of supply chain communications. A copy is available at http://www.nist.gov/director/prog-ofc/report04-2.pdf.

Process Specification Language (PSL) Becomes an International Standard

ISO 18629-1, Process Specification Language Part 1, "Overview and basic principles," was published as an international standard by ISO. NIST played an integral role in the initial drafting and continued leadership in the development of ISO 18629 and its associated parts. Process data is used throughout the life cycle of a product, from early in the manufacturing process during design, through process planning, validation, production scheduling, and control. The standard is available for purchase via the ISO online catalog:
http://www.iso.org/iso/en/ISOOnline.frontpage.

NIST B2B Testbed Demonstration with the Automotive Industry Action Group

NIST staff presented an update of the Object Application Group (OAG)/NIST B2B Testbed capabilities at the AUTOTECH 2004 conference in Detroit, MI. NIST presented the methodology developed for content testing that was applied to the Inventory Visibility and Interoperability (IV&I) project (http://www.aiag.org/whatsnew.cfm#ivi) as part of the IV&I status update, and demonstrated the testbed capability to support Business Object Document (BOD) constraint checking to the representatives of Standards in Automotive Retail (STAR) consortium. The STAR delegation expressed significant interest to use NIST capabilities for BOD management and testing of business rules and constraints within the STAR community. The NIST team also presented the vision and some technical ideas for a Semantic Web-based approach to develop the next generation OAG specification. For more information regarding the NIST test bed, please visit:
http://www.mel.nist.gov/msid/b2btestbed/.

FY2005 Projects

Interoperability Test Bed
(Objectives #1 & #2)

Combine existing and emerging MEL & NIST testing capabilities into a coherent, structured interoperability testing environment, with tools, test cases, test bed documentation; hold a demonstration of test bed capabilities at a public open house.

Virtual Manufacturing Environment
(Objectives #1 & #4)

Collect the manufacturing applications and simulation tools needed to support the various integration threads. Where possible, commercial applications will be acquired and training completed; otherwise, prototype software will be developed.

Supply Chain Business Systems
(Objectives #1 & #4)

Bring together the software applications and interface specifications that will provide capabilities to satisfy industry's priority needs in the integration of supply chain business systems.

Production Management Systems
(Objectives #1 & #4)

Bring together the software applications and interfaces specifications that will provide capabilities to satisfy industry's priority needs in the integration of production management systems.

Shop Process Systems
(Objectives #1 & #4)

Bring together the software applications and interfaces specifications that will provide capabilities to satisfy industry's priority needs in the integration of process engineering and shop floor hardware systems.

Collaborative Design and Engineering Systems
(Objectives #1 & #4)

Bring together the software applications and interfaces specifications that will provide capabilities to satisfy industry's priority needs in the integration of design and engineering systems.

Common Manufacturing Primitives Ontology Specification
(Objectives #1, #5, & #6)

Develop formal ontologies using emerging semantic methods such as Process Specification Language (PSL), Web Ontology Language (OWL) and Resource Description Framework (RDF), in support of the integration threads.

Next Generation Integration Technologies
(Objectives #1, #5, & #7)

Continue the work begun in the Automated Methods for Integrating Systems (AMIS) project developing tools, methods, and models, and using the ontologies developed for the integration threads, to support the automation of various aspects of the integration process.

Typical Customers and Collaborators

Industrial:
Accordare, Drake Certivo, Lockheed Martin, Nyamekye Research and Consulting Firm, Covisint, General Motor Corp., Ford Motor Company, Lear, Lesker Corporation, The Boeing Company, Deere & Company, LK Metrology, Mitutoyo, Pratt & Whitney, DaimlerChrysler, General Electric, LK, Zeis, and Nihon, Unisys

Consortium:
Automotive Industry Action Group, PDES, Inc., and Metrology Automation Association

Software Vendor:
AutoSimulation, Inc., EDS, Promodel Corporation, Micro Analysis & Design Incorporated, Softimage, Proplanner, Flexsim Software, Emergis, Fujitsu, QAD, SAP, Sterling Commerce, iConnect, Wolverine Software, Simul8 Corp., Delmia Corporation (formerly Deneb Robotics), Systems Modeling – Rockwell Software, Sewickley, Knowledge Based Systems, Inc., Technomatix, Delmia, Wilcox, and Theorem Solutions

FY2005 Standards Participation

Active participation and leadership roles in:

- ANSI/ASME B5 Machine Tools - Components, Elements, Performance, and Equipment
- Foundation for Intelligent Physical Agents ISO/IEC JTC1/SC22/WG11
- ISO/IEC JTC1/SC32 Information Technology Data Management and Interchange
- ISO TC184/SC4 Industrial automation/Industrial data
- ISO TC 184/SC1/WG7 Industrial automation systems and integration/ Physical device control
- ISO TC184/SC5/WG4 Manufacturing Programming Environments
- I++ Group
- OASIS
- Object Management Group
- Open Applications Group
- SISO/IEEE
- US PRO
- W3C Semantic Web

Programs of the Manufacturing Engineering Laboratory

Manufacturing Interoperability

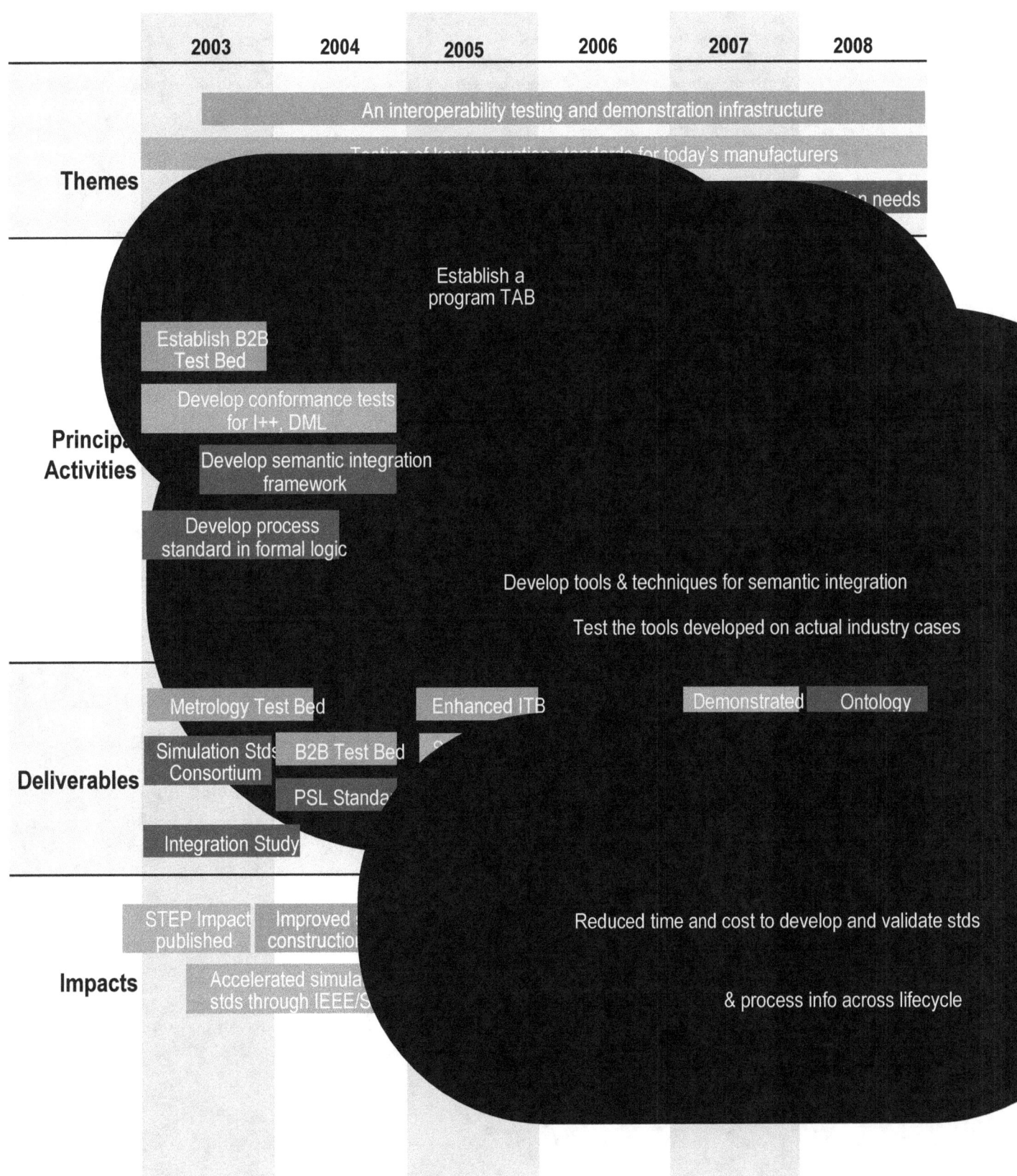

Programs of the Manufacturing Engineering Laboratory

Manufacturing Metrology and Standards for the Health Care Enterprise

Program Goal:

Apply proven MEL manufacturing technology and expertise to healthcare systems, biomedical devices and equipment, and biomedical data management

Program Manager:
Ram D. Sriram

Total FTEs:
2.4

Annual Program Funds:
$595 K

Customer Need & Intended Impact

Spending on healthcare in the United States was about 13.2 % of the Gross Domestic Product (GDP) in 2000, which is $1.3 trillion, and continues to grow at the rate of 7.3 % per year. This amount will reach $2.8 trillion dollars by 2011 (around 17 % of the GDP). These costs are also a major concern for the U.S. industry, as escalating healthcare costs are impeding our ability to compete globally. According to a USA Today article, General Motors (GM) spent $4.5 billion on healthcare in 2002, an increase of nearly 9 % from 2001. GM sold 8.4 million vehicles in 2002. In effect, healthcare expenses constituted $535 of the price tag of each GM car. This is reiterated in a recent U.S. Department of Commerce report entitled "Manufacturing in America: A Comprehensive Strategy to Address the Challenges to U.S. Manufacturers." This report cites that rising healthcare costs may prove to be detrimental to our manufacturing industry, with testimonials from various industries.

Healthcare and manufacturing share many similar organizational, technological and informational issues. Thus, the healthcare industry as a whole is a customer for the metrology, standard-setting support and technology approaches and solutions that MEL has developed for the manufacturing sector that are transferable or adaptable to the healthcare sector.

The benefit to the healthcare industry will be an infrastructure for the accelerated and enriched development of improved organizational, technological and informational support methodologies for all aspects of health care delivery. NIST's contributions will enable more effective development and application of biological and medical knowledge to practical problems.

Technical Approach & Program Objectives

There are two dimensions to the program: (1) Healthcare informatics; and (2) Medical devices. Healthcare informatics deals with all the processes or "software" of the healthcare enterprise: modeling and simulation, design and production, biosurveillance, manufacturing and its associated supply chains, and information and data management both in clinical practice and biological research. Medical devices deal with all the products or "hardware" of the enterprise: the characterization, design, manufacture, testing, and metrology of medical devices at scales ranging from large equipment to nano-scale drug delivery mechanisms.

This program deals with the following objectives:

(1) Healthcare informatics

Objective #1.1: Enterprise modeling and simulation
Explore the applicability of the modeling and simulation technologies developed in MEL to healthcare systems; explore means for disseminating this information to the shareholders in the healthcare industry.

Objective #1.2: Design and production of pharmaceuticals
Develop representations of pharmaceutical processes, quality measurement methods, and test equipment standards necessary for the specification, characterization, and data interchange involved in the clinical trials, certification, production testing, and manufacture of pharmaceutical products.

Objective #1.3: Biosurveillance
Develop information models and integration technologies, classifications of healthcare terminologies and ontologies, interchange specifications and test methods necessary for the acquisition, characterization, standardization, and validation of public health surveillance information and the dissemination and integration of relevant treatment guidelines to improve the detection and response to disease outbreaks and insidious bioterrorism attacks.

Objective #1.4: Manufacturing and value chain management
Develop interfacing specifications and interaction protocols for integrating manufacturing and e-commerce software solutions into the biomedical device value chain and enhance existing standards for device integration.

Objective #1.5: Clinical informatics
Extrapolate from MEL's experience in information modeling and research supporting information interchange standards development for the manufacturing industry to provide experience, assistance and leadership for related activities in the health care informatics field.

Objective #1.6: Bioinformatics
Adapt and extend NIST's expertise in information modeling, information interchange and standards development in the manufacturing arena to the field of bioinformatics, leading to synergisms with bioinformatics research and practice and consolidation of the bioinformatics knowledge base.

Programs of the Manufacturing Engineering Laboratory

(2) Medical devices

Objective #2.1: Mobility devices
Develop test methods and performance metrics, sensor data, standards and specifications necessary for intelligent assistive devices for wheelchair dependents and the blind.

Objective #2.2: Hearing devices
Develop test and measurement methods, data, standards and specifications necessary for the characterization, manufacturing, and testing of hearing devices and related diagnostic equipment.

Objective #2.3: Intelligent assistive surgical devices (medical robots)
Work with an American Standards for Testing of Materials (ASTM) committee, the Food and Drug Administration (FDA), medical robotic research groups and University Hospitals for the establishment of Intelligent Assistive Surgical Devices (Medical Robots) standards. This work will extrapolate on our previous work on industrial robot performance and safety standards, related metrology, instrumentation and artifact and marker design.

Objective #2.4: Surface characterization of biomedical devices
Develop test procedures for characterizing the surfaces of medical devices that relate to device function and failure behavior.

Acoustic research using KEMAR- a test manikin with sensors instead of eardrums

Objective #2.5: Meso-micro-biodevices
Assist in the establishment of meso-micro-biodevices standards. Meso scale devices have components with feature sizes of a few millimeters. Micro-scale devices have components with features, which range between 1 mm and 1 mm and nano-scale devices have components with features, which range between 1 μm and 1 nm.

Objective #2.6: Nano-biodevices
Develop protocols for high-resolution imaging of individual components and associated complexes of the constituents of nanoparticle drug delivery systems (NDS). Demonstrate imaging of the cell transfection process with fixed and live cells using such systems. Nano scale devices have components with features that range between 1 μm and 1 nm.

Major Accomplishments

This is a new program initiated in FY 2005. The program builds on the previous accomplishments of all five divisions of MEL, which include the following: an exploratory project on healthcare information interchange through shared ontologies; data representation schemes for proteomics standards; initial studies for robotic wheel chair standards; active participation in the ASTM Intelligent Assistive Surgical Devices (Medical Robots) standards meetings; an exploratory project on nanoparticle imaging which resulted in the acquisition of a scanning probe microscope with biological imaging capabilities; and a major role in the development of American National Standards Institute (ANSI) standard S3.22 "Specification of Hearing Aid Characteristics."

FY2005 Projects

A star (*) in front of a project indicates that the project will be pursued if we are successful in obtaining external funding/collaboration.

Simulation Applicability Study (Objective 1.1)

Prepare a comprehensive report on the applicability of modeling, simulation and visualization concepts developed at NIST/MEL to biological information and processes.

*Pharmaceuticals Industry Involvement Roadmap (Objective 1.2)

Develop a roadmap detailing potential tasks (e.g., similar to those described below) for NIST's role in aiding the pharmaceutical industry. Publish a workshop report on Interoperable Manufacturing Process Specifications to enable interoperability and interchange of manufacture of active ingredients and key intermediates in the production of pharmaceuticals.

Public Health Surveillance Data Interchange Proposal (Objective 1.3)

Develop a proposal for potential funding on standards for the interchange of public health surveillance data and integration of newly disseminated active clinical guidelines into healthcare information systems.

Biomedical Device Industry Needs Study (Objective 1.4)

Conduct a workshop to collect industry needs and prepare a thorough report on biomedical device manufacturers' needs for standards and a roadmap to their achievement. This report will also identify stakeholders, due dates, approaches, and appropriate standard activities for which we should participate.

Clinical Information Standards Plan (Objective 1.5)

Prepare a comprehensive report of all clinical information-oriented standards, their development organizations, their scope and the vocabularies/ontologies they employ. Use the report as the basis for developing a plan for applying NIST's experience to assist in clinical information-oriented standard development and closer harmonization.

*Bioinformatics Framework Proposal (Objective 1.6)

Prepare a comprehensive report reviewing bioinformatics concepts and languages used in bioinformatics, proposing a framework based on information modeling languages, formal ontologies and a common set of concepts enabling better communication.

Wheelchair Standards Core Competency Development (Objective 2.1)

Develop core competencies within MEL to address robotic wheelchair standards organizations intelligently. Prepare and submit a proposal to a healthcare funding organization to further this research.

Hearing Devices Standards Report (Objective 2.2)

Report on development and status of ANSI (The American National Standards Institute) and international (primarily IEC (International Electrotechnical Commission)) standards on hearing devices and related diagnostic equipment.

Standard Test Classes for Human Joint Specimens (Objective 2.3)

Select a few human joint specimens, which will be candidates for NIST standard reference artifacts. These specimens will come from healthy and diseased joints and will form separate standard test classes of the developing standard.

Biomedical Surface Characterization Workshop (Objective 2.4)

Plan and organize a workshop on surface characterization for the biomedical industry.

*Meso-Micro-Biodevice Technology Report (Objective 2.5)

Prepare a brief report describing the status of the Meso-Micro-Biodevices technology and possible opportunities for collaboration with other research groups.

Advanced Nanoparticle Imaging Opportunities (Objective 2.6)

Leverage external funding to demonstrate high-resolution imaging of individual components and associated complexes of the constituents of a NDS (Tf-Lip-p53).

Typical Customers and Collaborators

Healthcare providers and organizations
Cleveland Clinic, Kaiser Permanente, Union Hospital, MD, Mayo Clinic, Partners Healthcare, MA, and others; Pharmaceutical manufacturers such as Eli Lily and Company, Pfizer, Bayer, GlaxoSmithKline, Mylan Laboratories, etc.;

Process modeling vendors
Aspen Technologies; Healthcare informatics vendors and consultants, such as Apelon, Adam Inc., LightPhrama, Synergene Therapeutics, Inc., etc.; Medical device industry;

Academic institutions
University of Pennsylvania, Stanford Medical Informatics, Purdue University, University of Minnesota, University of Michigan, West Virginia University, Carnegie-Mellon University, Johns Hopkins University, University of Pittsburgh, University of Delaware, Georgetown University Medical Center, etc.;

Government organizations
Telemedicine and Advanced Technology Research Center, Fort Dietrick (MD), U.S. Army, U.S. Uniform Health Services, Department of Homeland Security, Various institutes of the National Institutes of Health, Food and Drug Administration, Centers for Disease Control, Veterans Administration, Agency for Healthcare Research and Quality;

Various associations
Radiological Society of North America (RSNA), ANSI, RESNA (Rehabilitation EFY2005 Standards Participation

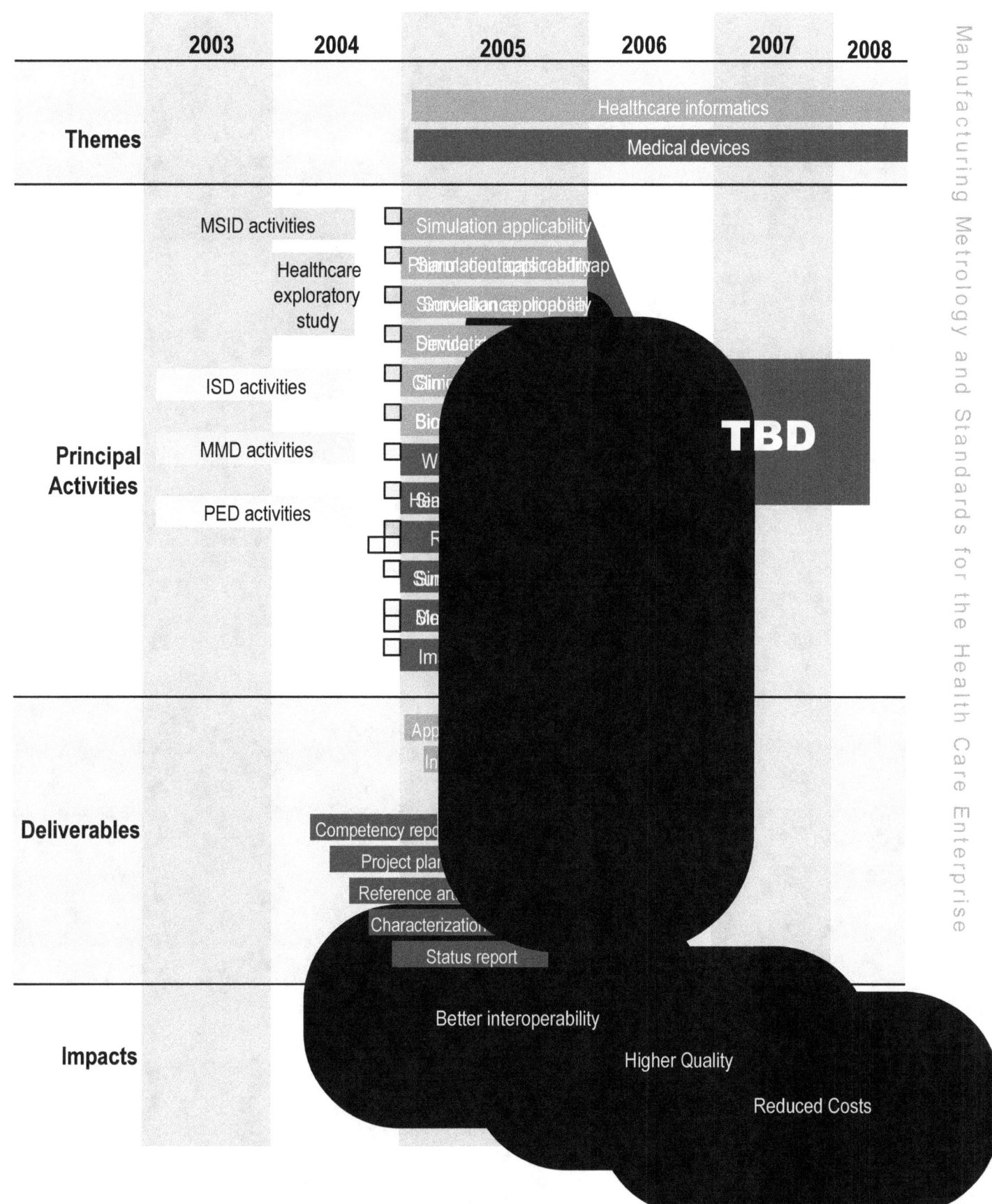

Mechanical Metrology

Program Goal

Develop and deliver timely measurements and standards to address identified critical U.S. industry for traceable mechanical metrology in the areas of acoustics, force, mass, and vibration, particularly for the support of trade and innovation, process-control, and quality in manufacturing.

Program Manager:
Zeina J. Jabbour

Total FTEs:
9.3

Annual Program Funds:
$2.327 M

Customer Need & Intended Impact

Mechanical metrology plays a critical role in nearly all sectors of the U.S. economy and in everyday life. The mechanical metrology program provides the research and development infrastructure for responding to and anticipating the needs of the U.S. government and industry in the areas of acoustics, force, mass, and vibration measurements. In addition, the program realizes, maintains, and disseminates the basic SI (System of International Units) unit of mass, and the quantities of acoustics, force, and vibration to a broad customer base covering the nuclear, aerospace, electronics, electric power, pharmaceutical, biotechnology, materials, chemical, construction, and automotive industries, state governments, as well as the U.S. departments of defense, labor, energy, transportation, veteran's affairs, and agriculture.

To insure the competitiveness of the U.S. industry in world markets, the mechanical metrology program maintains active participation in the CIPM (Comité International des Poids et Mesures) consultative committees on Mass and related quantities (CCM – Consultative Committee on Mass and Related Quantities) and Acoustics, Ultrasound, and Vibration (CCAUV – Consultative Committee on Acoustics, Ultrasound, and Vibration), the Interamerican Metrology System (SIM), participation in the Mutual Recognition Arrangement (MRA) and associated international key comparison. A crucial aspect of the mechanical metrology program is the transfer of NIST-developed measurement and uncertainty evaluation techniques to the U.S. industry and other government agencies. This is accomplished through participation and/or leading of standards activities (ISO, IEC - International Electrotechnical Commission, OIML - International Organization of Legal Metrology, ASTM – American Society for Testing and Materials) and other direct outreach activities that serve the dual purpose of assessing immediate and anticipated needs of the metrology and research and development (R&D) community as well as transferring the current resident knowledge and expertise at NIST.

The R&D needs include short-term and current needs that are the focus of the existing activities. It is the intention of this program to increase the emphasis on long-term anticipated needs. This will be accomplished over time in an evolutionary manner, beginning with forming high-level contacts with the R&D community to assess the impact of the current measurement services provided and to establish a roadmap for the long-term future of mechanical metrology.

Areas such as stable mass artifacts, traceable vacuum mass measurements, optical detection of forces, MEMS (Micro Electro-Mechanical Systems) load cells, shock metrology, traceable phase measurements for accelerometers, next-generation measurement techniques for new microphones, and angular accelerometers are areas of known interest for industry and public-sector customers would result in significant anticipated impacts cutting across most sectors of the U.S. economy. The short-term/current needs are evidenced by the large demand for the measurement services provided and the continuous requests to improve and upgrade, existing services, and/or develop new ones. Such additional needs include the development of low-frequency vibration capabilities, improved density measurements, torque, dynamic force, and dynamic torque measurements.

Technical Approach & Program Objectives

The program goal will be met through prioritization of activities and allocation of resources to address the key elements of each mechanical metrology service area. For each service area, the activities of R&D for new and improved measurement capabilities, provision of services to customers, participation in international comparisons to eliminate barriers to trade, and standardization of measurement methods and protocols are integral to the overall success of the program. These aspects form the basis for the program objectives and each objective interacts strongly with the others. The program will draw upon the internationally recognized skills and expertise of MEL staff in the mechanical metrology areas to achieve the program objectives. Funding will be leveraged from multiple sources as much as possible to enable appropriate allocation of resources. As indicated above, the program intends to increase the emphasis on R&D to respond to customer needs for future measurement capabilities and to increase the interactions with mechanical metrology R&D organizations and the "end-users" of devices and artifacts calibrated by MEL measurement services.

Objective #1: Research and Development Activities

By FY2009, conduct research and development activities to anticipate the long-term, short-term, and current needs of fundamental SI metrology, government, and industry, and drive future advancements in the areas of mass, acoustics, force, and vibrations.

Deliverables for FY2009:

Foundation for "future mass metrology" built by designing, building, and testing a prototype magnetic levitation vacuum balance, and developing stable mass artifacts to provide the foundation for the first directly-traceable vacuum measurements, eliminate the time-variance of the mass artifacts and the definition of the Kilogram and enables the realization of the alternative definitions of the artifact mass definition.

Milestones:

- For 2005, preliminary design of balance including weighing mechanism, vacuum/air interface chamber, and magnetic levitation and shielding mechanisms.

- For 2007, the magnetic rigid link system and report on the operation of the balance in air and in vacuum.

- For 2007, mechanisms for in-situ measurements of surface and material properties.

- For 2007, methods for manufacturing stable mass artifacts and evaluation of platinum-iridium and stainless steel samples.

- For 2008, the first set of stable mass standards and report on surface and material properties monitoring.

- For 2009, full implementation of first prototype magnetic levitation balance; first vacuum/air mass measurement with direct traceability to one of the U.S. national prototype Kilograms; and a report of operation uncertainty sources, and requirements for building the next-generation balance to realize measurements with 1×10^{-10} relative precision.

Deliverables for FY2006:

Robotic mass and solid density measurement facility built in the Advanced Measurement Laboratory (AML) to eliminate errors caused by human interface and optimize measurement process. This facility will cover the mass range from 1 mg to 64 kg and solid density of artifacts in the range from 10 g to 1 kg and will be used to improve the measurements efficiency and uncertainty of NIST reference standards used in R&D and those used in delivery of measurement services. This facility will be adapted for the provision of the mass calibrations in the ranges specified and all procedures and uncertainties documented.

Milestones:

- For 2005, implementation of mass and density robotic systems in AML. and measurement procedures for robotic systems.

- For 2006, robotic systems for the provision of mass and solid density measurement services and procedures documented.

Programs of the Manufacturing Engineering Laboratory

Deliverables for 2005:
The automation and upgrade of hardware and software controls of the 27 kN (6.1 klbf) force deadweight machine completed to allow hysteresis measurements, enhance efficiency and improve the consistency of the measurement process and document all operation procedures, hardware, software, and testing results in NISTIR.

Deliverables for 2005:
The extended low-frequency sinusoidal accelerometer calibration system with improved accuracy and extended frequency range to satisfy the requirements of the automotive, aerospace, construction, nuclear power, manufacturing industries, and Department of Defense (DoD) for traceability of vibration transducers in the frequency range extending at least an order of magnitude below 1 Hz; documentation of developed procedures and uncertainty analysis necessary for establishing a new NIST SP250 calibration service; and vibration measurement services implemented in new laboratory in AML.

Objective #2: Customer Interactions
By 2008, establish and/or maintain contacts with customers from various sectors of the U.S. industry, government agencies and departments, and academia to evaluate and assess the future research and development needs in the areas of mechanical metrology.

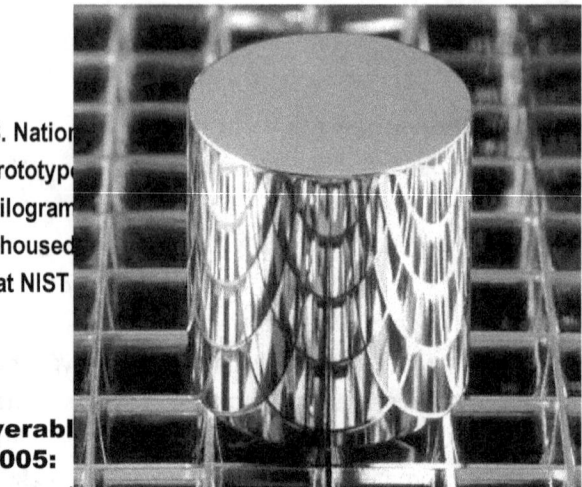

U.S. National Prototype Kilogram housed at NIST

Deliverables for 2005:
- Major mechanical metrology customers identified and contacts initiated with assistance from MEL strategic manager.
- Site visits to selected customers to identify the current impact of the mechanical metrology services and identify the future needs; and a summary report of findings.

Deliverables for 2006:
Multiples workshops to formally assess and identify the industry needs in the acoustics, force, mass, and vibration and reports on workshops results and findings.

Deliverables for 2008:
Consensus technology roadmaps developed with industry contacts, that to plan and prioritize development of future mechanical metrology capabilities based on identified industry needs and workshop results in mass, acoustics, force, and vibration.

Objective #3: Measurement Services

Provide measurement services in conformance with an ISO 17025 compliant quality system to meet the needs of the U.S. government and industry in the areas of acoustics, force, mass, and vibrations. Activities associated with this objective are ongoing with annual deliverables.

Annual Deliverables:

1. Provided SP250 calibration services and special tests to broad customer base in the areas of acoustics, force, mass, and vibration with uncertainties equal to or better than those specified in publication NIST SP 250; and 90 % or better on-time delivery of reports of calibrations and special tests to all customers.

2. Compliance with ISO 17025 quality systems for all SP250 measurement services from all administrative and technical aspects including regular calibration of NIST reference and working standards, calibration of all equipment and/or instrumentation associated with calibration systems with direct traceability to NIST, and maintaining control charts. Report on status to Measurement Services Advisory Group (MSAG).

3. All calibration fees reviewed and revised as necessary to insure full cost recovery for all calibration services in accordance with NIST policies.

Objective #4: International Key Comparisons

By 2008, conduct international key comparisons to insure U.S. compliance with the MRA and insure competitiveness of the U.S. industry in world markets.

Deliverables for 2008:

Participate and/or pilot, perform measurements, and generate reports for the CIPM and SIM key comparisons in acoustics:

Milestones:

- For 2005, participation in the CCAUV comparison of laboratory standard microphones at 31.5 Hz to 25 kHz; analysis and comments on Draft A results.
- For 2006, participation in the CCAUV comparison of free field sound pressure in air in the range of 2 kHz to 40 kHz (if enough other National Measurement Institutes (NMIs) are ready),
- For 2006, participation in the CCAUV comparison of laboratory standard microphones at low frequencies in the range of 2 Hz to 125 Hz.
- For 2008, participation in the SIM comparison of laboratory standard microphones in the range of 125 Hz to 8 kHz.

Programs of the Manufacturing Engineering Laboratory

Deliverables for 2008:
Participate and/or pilot, perform measurements, and generate reports for the CIPM and SIM key comparisons in force:

Milestones
- For 2005, pilot lab for the CCM 2 MN and 4 MN comparisons.
- For 2006, participation in the CCM 10 kN comparison.
- For 007, participation in the CCM 100 kN comparison.
- For 2008, participation in the CCM 1 MN comparison.

Deliverables for 2008:
Participate and/or pilot, perform measurements, and generate reports for the CIPM and SIM key comparisons in mass:

Milestones
- For 2005, co-pilot lab for the CCM comparison of 1 kg stainless steel artifacts.
- For 2006, participation in the CCM comparison of solid density.
- For 2006, participation in the CCM comparison of 50 kg mass standards.
- For 2008, participation in the CCM comparison of multiples and submultiples of 1 kg standards.

Deliverable:
Participate and/or pilot, perform measurements, and generate reports for the CIPM and SIM key comparisons in vibration:

Milestones
- For 2005, pilot lab for the SIM comparison in the range of 50 Hz to 5 kHz.
- For 2006, participation in the SIM low frequency vibration comparison.

Objective #5: Standards Activities

Provide high-level technical expertise through participation in standards activities in the areas of acoustics, force, mass, and vibration to ensure traceability and comparability of U.S. physical and documentary standards in mechanical metrology to those of other nations and to represent and protect the interests of the U.S. industry in global markets. A complete list and roles of mechanical metrology program staff in standards committees is listed under the "standards participation" section. Activities associated with this objective are ongoing with annual deliverables.

Deliverables:
NIST interests represented and defended contribution to future planning activities, and status reports on technical developments and MRA related activities at the meetings of the CCM and associated working groups on mass, force, and density in the years 2005, 2006, 2008, and 2009, CCAUV in the years 2006 and 2008, SIM and its associated working groups on mass and related quantities, and acoustics, ultrasonics, and vibrations as needed.

Annual Deliverables:
As Technical Advisor to the USNC (ANSI's United States National Committee)/IEC for Technical Committee (TC) 29, recommendation USNC/IEC position and contribution to the development of acoustics standards (annual deliverables):

- IEC 61094-6 "Measurement microphones – Part 6: Electrostatic actuators for determination of frequency response,"

- IEC 61094-7 "Measurement microphones – Part 7: Values for the difference between free-field and pressure sensitivity levels of laboratory standard microphones,"
- IEC 61672-3 "Electroacoustics – Sound level meters – Part 3: Periodic tests,"
- Drafts/proposals for new ANSI standard S1.15 Measurement Microphones – Part 2: Primary Method for Pressure Calibration of Laboratory Standard Microphones by the Reciprocity Technique, for other parts of the S1.15 series, and for new ANSI standards on sound calibrators, on sound level meters.

Deliverables for 2005:
- Publication of ANSI S2.1 on Vibration and Shock Terminology.
- Publications of ISO 16063-15 on Primary Angular Vibration Calibration by Laser Interferometry and ISO 16063-22 on Shock Calibration by Comparison to a Reference Transducer.

Deliverables for 2008:
Publications of ISO 18431-1 on a General Introduction to Mechanical Vibration and Shock Signal Processing in 2007 and ISO 18431-4 on Shock Response Spectrum Analysis in 2008.

Deliverables for 2009:
Revision of ISO 2041 (TC108 WG1) on Vibration and Shock Terminology and integration of ANSI S2.1 into it.

Deliverables:
- For 2005, technical contributions to resolve transducers stability criteria in ASTM E74-02 "Standard Practice of Calibration of Force-Measuring Instruments for Verifying the Force Indication of Testing Machines" and revised standard.
- Annual review the ASTM Force Standards E74-02 and E4-03 "Standard Practices for Force Verification of Testing Machines" and participation in bi-annual committees / sub-committees meetings as needed.

Deliverable
Technical contributions to OIML R111 and ASTM E617-97 standards on weights, their classifications, technical requirements, and measurement procedures as needed.

Programs of the Manufacturing Engineering Laboratory

Major Accomplishments

ISO 17025 compliant quality system implementation and assessment for international recognition of services and participation in the MRA

- Development, documentation, implementation, and assessment of the MEL Manufacturing Metrology Division (MMD) quality system for measurement services in Acoustics, Force, Mass and Vibration completed.

- Quality system assessed both internally by MMD and by NIST at large and found to be in conformance with the NIST quality system by the NIST Assessment Review Board and the NIST Measurement Services Advisory Group.

- Electronic copy of quality manual and report on quality system submitted to SIM Quality System Task Force for approval during November 2004 Regional Board meeting. This quality system supports the NIST Calibration and Measurement Capabilities claimed for Acoustics, Force, Mass and Vibration in Appendix C of the Mutual Recognition Arrangement.

Strong support for MRA through international and regional intercomparisons to eliminate trade barriers for customers

- Hosted meeting of SIM Metrology Working Group 9 on Acoustics, Ultrasound & Vibration.

- Draft A of the report on CCAUV.A-K3 prepared & circulated among the participating labs.

- Piloted Force 4 MN key comparison, CCM.F-K4; status report submitted for March meeting of CCM Working Group on Force.

- Final reports of results of key comparisons in mass, CCM.M-K1 and CCM.M-K2 published in Appendix B of the BIPM (Bureau International des Poids et Mesures) key comparison data base.

High quality NIST SP 250 measurement services

- In 2004, 442 calibrations, special tests, and NTEP (National Type Evaluation Program) tests were performed for 98 distinct customers in the industrial, governmental, and educational sectors.

Crucial role in development and publication of documentary standards to support measurement methodology and their implementation in calibration procedures

- ISO standard on secondary calibration of vibration and shock transducers by comparison published.

- ISO standard on time domain windows for Fourier transform analysis submitted for publication.

- IEC/TS 62370 on electromagnetic and electrostatic compatibility requirements and test procedures for sound-intensity measuring instruments published in May 2004.

- Four national standards published: ANSI S1.11-2004 on octave-band and fractional octave-band analog and digital filters, ANSI S1.17-2004/Part1 on measurements and specification of insertion loss of microphone windscreens, ANSI S3.6-2004 on specifications for audiometers, and ANSI S3.21-2004 on methods for manual pure-tone threshold audiometry.

FY2005 Projects

Mass Metrology of the Future (Objective #1)

Develop preliminary design of a vacuum balance that directly links the vacuum unit of mass to the current realization in air and begin manufacturing and/or procuring parts. Implement robotic mass facility in AML for mass and solid density metrology.

Upgrades and Expansion of Mechanical Metrology Measurement Services (Objective #1)

Expand the vibration measurement services in the low-frequency range. Implement new vibration labs in AML. Automate the 27 kN (6.1 klbf) deadweight machine to improve operation and realization of the unit of force.

Outreach (Objective #2)

Identify major mechanical metrology customers and initiate contacts with assistance from MEL strategic manager. Organize and conduct site visits to selected customers to identify the current impact of the mechanical metrology services and identify the future needs. Generate a summary report of findings.

Mechanical Metrology Measurement Services (Objective #3)

Provide measurement services in the areas of acoustics, force, mass, and vibration in accordance with the NIST quality system. Maintain the quality manual in compliance with ISO 17025 and NIST QMI (Quality Manual I). Update manual as needed. Issue quarterly reports to the MSAG on the status of the quality system for the areas in mechanical metrology.

International Key Comparisons and Standards Activities (Objectives #4 & #5)

Complete measurements and analyze results of the 4 MN key comparisons piloted by NIST in collaboration with the NIST Statistical Engineering Division. Write and submit first draft report to the CCM working group on Force. Pilot SIM key comparison in vibration; conduct measurements, analyze results, and issue preliminary report. Co-pilot 1 kg mass standards key comparison with BIPM. Participate in CCAUV key comparison in acoustics. Represent NIST at the CCM meeting and the meetings of the Working Groups on Mass and Density. Represent NIST and the U.S. in meetings of the ASTM, ISO, IEC, OIML, and other standards organizations as needed.

Programs of the Manufacturing Engineering Laboratory

Typical Customers and Collaborators

Aerospace industry, automotive industry, construction industry, nuclear power industry, pharmaceutical industry, instrument manufacturers, university research labs, state weights and measures labs, federal agencies (Departments of Agriculture, Commerce, Defense, Energy, Labor, Veterans Affairs, Justice, and the Food and Drug Administration).

FY2005 Standards Participation

ANSI S1 Acoustics:
Organizational Member; S1/WG1 Standard Microphones and their Calibration: Chair; S1/WG21 Electromagnetic Susceptibility of Acoustical Instruments: Member

ANSI S2 Mechanical Vibration and Shock:
Vice Chair and Organizational Member; S2/WG2 Terminology: Chair; S2/WG3 Signal Processing Methods: Member; S2/WG5 Use and Calibration of Vibration and Shock Measuring Instruments: Chair

ASACOS (Acoustical Society of America on Standards):
Member

ASTM E7 Nondestructive Testing:
Member; E7.04 Acoustic Emission: Member; E7.06 Ultrasonics: Member.

ASTM E28 Mechanical Testing:
Member; E28.01 Calibration of Mechanical Testing Machines and Apparatus: Member; E28.05 Residual Stress: Member; E28.13 Dynamic Modulus Measurements: Member.

ASTM E41.06 Weighing Devices:
Member

CIPM:
Consultative Committee on Acoustics, Ultrasound and Vibration (CCAUV): Delegate; Consultative Committee on Mass and Related Quantities (including force) (CCM): Delegate; CCM WG on Mass: Member; CCM WG on Density: Member; CCM WG on Force: Member.

SIM:
MWG7 Mass & Related Quantities, MWG9 Acoustics and Vibration: Member.

IEC TC29 Electroacoustics:
Chair U.S. Delegation; USNC/IEC for TC29: Technical Advisor; TC29/WG5 Measurement Microphones: Member.

ISO TC108 Mechanical Vibration and Shock:
Chair U.S. Delegation; TC108 U.S. TAG: Chair; TC108/WG1 Terminology: Member; TC108/WG26 Signal Processing Methods for the Analysis of Mechanical Vibration and Shock: Convener; TC108/SC3 Use and Calibration of Vibration and Shock Instrumentation: Chair U.S. Delegation; TC108/SC 3 U.S. TAG (Technical Advisory Group): Chair; TC108/SC3/WG6 Calibration of Vibration and Shock Transducers: Expert Member; TC108/SC3/WG10 Vibration Condition Monitoring Transducers and Instrumentation: Member.

ISO TC135 Nondestructive Testing:
Alternate Chair U.S. Delegation; ISO TC135/SC3 Acoustic Methods: Chair; ISO TC135/SC3: liaison to IEC TC87 Ultrasonics.

Mechanical Metrology

Programs of the Manufacturing Engineering Laboratory

Mechanical Metrology

OIML TC9 Instruments for Measuring Mass and Density:
Technical Advisor to U.S. Voting Member; OIML TC9/SC3 Weights: Technical Advisor to U.S. Voting Member; USNWG/OIML TC9/WG1 Load Cells: Member.

OIML TC13 Measuring Instruments for Acoustics and Vibration:
Technical Advisor to U.S. Voting Member

FY2005 Measurement Services

Calibrations and Special Tests:
Provide calibrations and special tests as described under Mass Standards, Force Measurements, Vibration Measurements, Acoustic Measurements, and Ultrasonic Measurements in the NIST Calibration Services Users Guide, SP 250.

Research Facilities:
Provide tests and measurements on an as-needed basis in special NIST Research Facilities such as the NIST Acoustic Anechoic Chamber.

Testing:
Load Cell Evaluation - Provide evaluations of prototype load cells in accordance with both national (NTEP) and international (OIML R60) standards.

Army helicopter loudspeaker unit setup for testing in the acoustic anechoic chamber research facility

Programs of the Manufacturing Engineering Laboratory

Mechanical Metrology

- Enhance delivery of measurement services
- ...m outreach and assess present and future customer needs
- Provide calibrations services and special tests
- Establish R&D foundation for program
- CCAUV, SIM MWG7 & SIM MWG9

- Robotic mass facility
- Site visits, workshops, and roadmaps for R&D

- Access to state-of-the-art traceable measurements and standards
- Equivalency of measurements and open world markets

Nanomanufacturing

Program Goal

Develop and deliver timely measurements, standards, and infrastructural technologies that address identified critical U.S. industry and other government agency needs for innovation and traceable metrology, process-control and quality in manufacturing at the nanoscale.

Program Manager:
Michael T. Postek

Total FTEs:
25

Annual Program Funds:
$5.029 M

Customer Need & Intended Impact

Advanced nanomanufacturing is key to the strength and future growth of the U.S. manufacturing sector and a strong measurements and standards infrastructure is vital for its success. NIST is responsible to U. S. manufacturing for providing traceability to the national unit of length, by developing measurement capabilities and calibration standards. Metrology is a key enabler for all manufacturing and it is especially important to nanotechnology. It has been predicted that within the next 10 years, at least half of the newly designed advanced materials and manufacturing processes will be built at the nanoscale. Measurement science (metrology) and advanced instrumentation are essential for nanomanufacturing. If you cannot measure it you cannot make it and that statement is even more accurate in the regime of nanotechnology. Successful metrology infrastructure is essential for manufacturers to achieve the real promise of developing and manufacturing new nanomaterials, devices, and products. Advanced instrumentation provides the necessary data upon which sound scientific conclusions can be based and correct metrology provides the ability to properly and accurately interpret those data. Together they facilitate nanomanufacturing. As pointed out in the National Nanotechnology Initiative (NNI) Instrumentation and Metrology Grand Challenge workshop final report, some of these metrology techniques will be evolutionary and some will be revolutionary. With that in mind, it is imperative that this program remain agile and evolve with the nanomanufacturing industry and adapt as new applications develop. It has been proven that where the economy of scale in manufacturing is concerned, even relatively small improvements in the metrology in the manufacturing process can yield large savings and increased value to the U. S. Economy.

Technical Approach & Program Objectives

This program represents an integrated approach to the currently known nanomanufacturing needs with an eye on the anticipated needs which will soon emerge. Therefore agility will be the key to the success of the technical approach. By integrating instrument development, and metrology infrastructure development a strong package of MEL metrology capability can be presented to the customer base.

Objective #1: SEM for Nanoscale Measurements

Develop the best-in-the-world accurate 3-dimensional Scanning Electron Microscope (SEM)-based dimensional metrology and modeling capable of measurement resolution of less than 0.1 nm and apply these to real-world samples to meet the requirements of the current and emerging nanomanufacturing in microelectronics, nanotechnology and biotechnology.

Deliverables for FY2005:
The new high-resolution, environmental, laser interferometer stage metrology SEM installed in the Advanced Measurement Laboratory (AML).

Deliverables for FY2006:
Extension of modeling capabilities of NIST's Monte Carlo (MONSEL) electron trajectory simulator to include two- and three-dimensional samples of arbitrary shape.

Deliverables for FY2007:
New, accurate 3-dimensional metrology coupled with modeling that allows the comprehensive experimentation on both virtual and real samples with nanometer uncertainty.

Deliverables for FY2008:
Validated MONSEL for low electron landing energy images and its capabilities extended to simulate low-vacuum (environmental) imaging modes (used for charge compensation when imaging insulating samples such as photomasks or biological samples) and the results published

Deliverables for FY2009:
A new state-of-the-art reference dimensional metrology SEM and accurate metrology methods for full-size wafers and masks. This work will carried out in cooperation with participants from the U.S. nanomanufacturing industry.

Objective #2: Optical Metrology

Provide world class metrology capabilities and technical leadership using optical based methods capable of measuring photomasks, reflection mode targets (as on wafers or industrially relevant substrates), and feature to feature positions with unsurpassed accuracy on features used in the most advanced nanoelectronics manufacturing.

Deliverables for FY2005:
Models and metrology targets; new overlay metrology methods that support the manufacturing of nanolelectronic devices at the 65 nm node and that can extend optical technology to the 65 nm sized manufacturing domain.

Deliverables for FY2006:
New scattering models for line width evaluation in transmission mode; advanced models necessary for optical metrology at the 32 nm node including reflection and transmission modeling of phase shifting features developed and tested in conjunction with industry.

Deliverables for FY2007:
A comprehensive set of Overlay Wafer Standard Reference Materials (SRMs), calibrated photomask linewidth features and 2-dimensional grids artifacts.

Deliverables for FY2008:
Advanced optical measurement methods capable of sub-1 nm repeatability and nanometer accuracy for use in nanoelectronics and nanomanufacturing feature metrology and process control.

Deliverables for FY2009:
A new optical metrology methodology useful in integrated metrology applications based on scatterfield microscopy.

Objective #3: Atom Based Metrology

Develop the best in the world atomic-scale metrology for measurements and standards in support of nanomanufacturing and nanoelectronics. Enable the capability to measure feature dimensions and positions with atomic precision and to fabricate test structures with sub-5 nm dimensions to support and enable developing advanced atomic-scale measurement capabilities.

Deliverables for FY2005:
Procurement of a new Scanning Tunneling Microscope (STM) that has improved atomic scale imaging capability for atom-based dimensional metrology.

Deliverables for FY2006:
Improved methods for etching nanostructures written in silicon work (in collaboration with ISMT (International Sematech) and the NIST Electronics and Electrical Engineering Laboratory (EEEL).

Deliverables for FY2007:
Techniques for the preparation of Scanning Probe Microscope (SPM) tips with reproducible geometries and the direct characterization of the SPM tip geometry and dimensions on the atomic scale.

Deliverables for FY2008:
Evaluation and demonstration of the use of nanotube tips for use in atomic scale metrology and nanolithography. Perform field emission testing to evaluate electrical characteristics.

Deliverables for FY2009:
Features written in silicon with critical dimensions smaller than 3 nm; The process developed so it can be implemented in other nanolithography systems and measured in external metrology systems.

Objective #4: SPM for Nanoscale Measurements

Develop world class traceable calibrations of probe-based calibration systems and procedures for the measurement of dimensional parameters for the semiconductor and other microelectronics industries with nanometer- and subnanometer-level uncertainties.

Deliverables for FY2005:

A SPIE (International Society for Optical Engineering) presentation and article on the development of the Veeco SXM (Scanning Probe Microscope) measurement system for linewidth measurement.

Deliverables for FY2006:

Calibrated pitch, height, and linewidth artifacts for maintaining and demonstrating the traceability of the instrument and SPIE presentation and article on the system concept and measurement results.

Deliverables for FY2007:

An independently traceable linewidth measurement technique, based on image stitching and the use of nanotube probes, for independent verification of the critical dimension (CD)-Atomic Force Microscope (CD-AFM) and an optimized reference measurement system at ISMT closely coupled both to NIST and to industry, providing traceable measurements of pitch, step-height, linewidth, and Line-Edge Roughness (LER).

Deliverables for FY2008:

A journal article on the comparison of the results of three independently traceable linewidth measurement techniques: the CD-

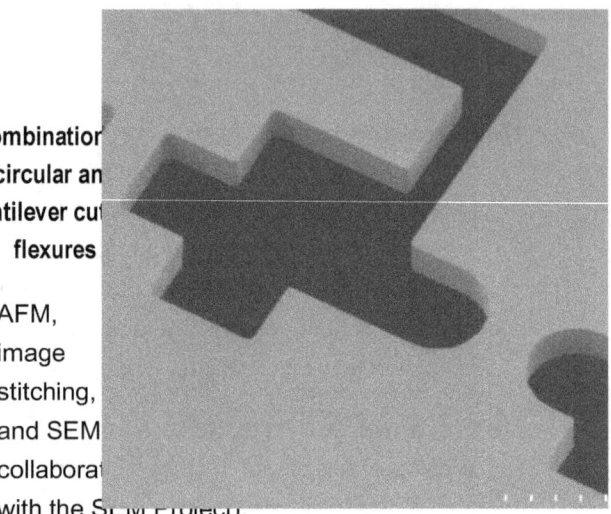

Combination circular and cantilever cut flexures

AFM, image stitching, and SEM collaboration with the SEM Project).

Deliverables for FY2009:

Traceable, CD-AFM-based linewidth measurement service for linewidth as small as 50 nm with uncertainties consistent with the sub nanometer resolution levels now required in industrial measurements (see: the International Technology Roadmap for Semiconductors ITRS) and capable of pitch and step height measurements as well.

Objective #5: Force Metrology for Nanoscale Measurements and Standards

Develop world-class force metrology for nanoscale measurements and standards for nanomanufacturing capable of improving the realization of the unit of force below 10^{-5} N by measuring a nanonewton with an expanded relative uncertainty below a percent.

Deliverables for FY2005:

A paper and review article on microforce submitted to the SPIE Microlithography Conference to introduce this topic to that community with potential resubmission to the Journal of Microlithography, Microfabrication and Microsystems.

Deliverables for FY2006:
A force and displacement measuring instrument capable of moving a rigid probe element or load cell into contact with the moving pan of the Electrostatic Force Balance (EFB) while recording force with 10 pN resolution and observing the relative separation of the probe and balance pan with 10 pm resolution.

Deliverables for FY2008:
A secondary force standard based on piezoresistive load sensing.

Deliverables for FY2009:
Examination of intrinsic force standards.

Objective #6: Advanced Control Systems and Positioning

Develop world-class advanced control and positioning systems for nanoscale measurements, assembly and standards. This includes arrays of high-precision Micro Electro-Mechanical Systems (MEMS) stages for scanning probe microscopy; multi-degree-of-freedom nanometer resolution fiber optic displacement sensors; and the automated assembly of micro and nano-scale components.

Deliverables for FY2006:
Fabrication of a 2 x 2 array of MEMS-based x-y-z positioning stages. The displacement range, frequency bandwidth and parasitic errors determined using an SEM. Additional tests will be performed at the EEEL MEMS testing laboratory to determine the thermal properties and out-of-plane stage motion. The results presented at an appropriate scientific conference and submit a paper to an appropriate scientific journal.

Deliverables for FY2006:
Integration of existing high-precision micromanipulation robot with an intelligent control architecture. This system will be used to prototype various microassembly operations that are critical to industry needs. Demonstration of a 4 degrees-of-freedom (DOF) peg-in-hole microassembly operation using NIST-developed microcomponent artifacts.

The results presented at an appropriate scientific conference and submit a paper to an appropriate scientific journal.

Deliverables for FY2007:
Test arrays of high-precision MEMS stages for scanning probe microscopy that will have better accuracy and resolution than commercial instruments and that will provide increased scanning area per unit of time due to parallel operation; The results presented at an appropriate scientific conference and submit a paper to an appropriate scientific journal.

Deliverables for FY2007:
A calibrated six degree-of-freedom fiber optic displacement sensor. It will be incorporated into a scanning probe microscope to measure the displacement and rotation of the scanning probe. The results presented at an appropriate scientific conference and submit a paper to an appropriate scientific journal.

Objective #7: Optical Tweezers for Nanoscale Manipulation and Metrology

Develop a best in the world capability to manipulate, assemble and test nano-scale devices such as nanowires using optical forces that combine laser-based manipulation, operator interface and automation to open new avenues for creating and testing nanodevices.

Deliverables for FY2005:
Functional nanodevices developed using semiconductor nanowires and publication of the results in a leading journal such as Applied Physics Letters.

Deliverables for FY2006:
A particle count standard fabricated by assembling a fixed number of particles into a suitable sample for calibration of particle counting instruments (in liquid) and disseminate to appropriate alpha test sites.

Deliverables for FY2009:
A new Nano-tweezers instrument based on innovative trapping physics and imaging techniques to manipulate and visualize much smaller components than is currently possible anywhere in the world using grid computing to allow 3D visualization in real time with resolution better than 30 nm.

Objective #8: Advanced Lithography

Develop a world class capability to fabricate test samples and standards using scanning probe oxidation and utilizing the unique capabilities afforded by the high accuracy scanning probe placement of the Molecular Measuring Machine; and develop a competence in imprint lithography which will enable nanomanufacturing of standards.

Deliverables for FY2005:
Reliable, effective methods and parameters for pattern transfer (i.e., etching) that utilize the SPM oxidation features as a mask.

Deliverables for FY2006:
A predictive model for SPM oxidation kinetics for optimized line width control of latent oxide features. The model will include the influence of electronic and ionic transport on the intrinsic thickness growth and lateral spreading due to space charge.

Deliverables for FY2007:
Prototypes of 1-D and 2-D silicon calibration structures.

Deliverables for FY2008:
Replication of sub-50 nm features using nanoimprint lithography, and verification of the fidelity and scale accuracy of the transferred patterns over macroscopic distances

Deliverables for FY2009:
A photomask for a standardized template for directed cell growth.

Objective #9: Nanomachining Technologies

Develop enabling technologies for Nanomanufacturing utilizing machine technologies to build interfaces that link nano devices to the physical world as well a fabricate imprint lithographic masks.

Deliverables for FY2005:
Participation in World Technology Evaluation Center (WTEC) survey of state of the art in micro machining.

Deliverables for FY2006:
Journal paper detailing design and characterization of metrology frame for improving accuracy of desktop machine used for micro machining.

Deliverables for FY2007:
Journal paper detailing micro/nano imprint manufacturing methods developed.

Deliverables for FY2009:
Demonstration of the combination of machining methodology with lithographic capabilities for nanoimprint masks.

Major Accomplishments

NNI Workshop on Instrumentation and Metrology for Nanotechnology

NIST held a very successful workshop on Instrumentation and Metrology for Nanotechnology in support of the National Nanotechnology Initiative. More than 250 attendees participated in the plenary sessions as well as the five breakout tracks. The overall conference was organized and chaired by Michael Postek and the Nanofabrication Breakout session was co-chaired by Richard Silver. There were numerous presentations in the sessions by international leaders within the various nanotechnology specialties. The output goal is a document that summarizes key goals, obstacles, roadblocks, and strategies aimed at achieving the visionary goals. This document will be used by the President's Council of Advisors on Science and Technology (PCAST) and other government funding agencies to provide direction and need for key research objectives and collaborations.

Successful Completion Of SEMATECH Assignment

Ronald Dixson, a program member, completed a two year assignment at International SEMATECH. The primary focus of the assignment was the development of a reference measurement system (RMS) using a critical dimension atomic force microscope (CD-AFM). The Veeco Dimension X3D – which is the current commercially available CD-AFM – was implemented as an RMS. Uncertainty budgets were developed and the instrument performance was thoroughly characterized.

Comparison of AFM-based Line Edge Roughness (LER) Measurements

A comparison of AFM-based LE[R] measurements were performed using the SEMATECH Advance[d] Metrology Advisory Group's pro[totype] standards with an intentionally c[oarse] LER. Different probes were use[d including] carbon nanotube tips. The stud[y showed] the carbon nanotube AFM tips p[erformed] better than other techniques both in terms of the ability to image closely spaced features and in being able to reach deep valleys.

Scanning Electron Microscope Modeling Proposal and Line Edge Roughness Recommendations

The meetings of the International SEMATECH (ISMT) Metrology Council draw a mix of metrologists representing ISMT member companies (semiconductor manufacturers), manufacturers of metrology tools, and government, business, and academic researchers. John Villarrubia, a program member, made a proposal for development of scanning electron microscope (SEM) models to support measurement of a greater range of sample shapes and compositions with more general detector placement. John also described two possible solutions to a measurement bias problem in line edge roughness measurement and recommended changes in the industry roadmap, since current specifications actually favor the incorrect measurement.

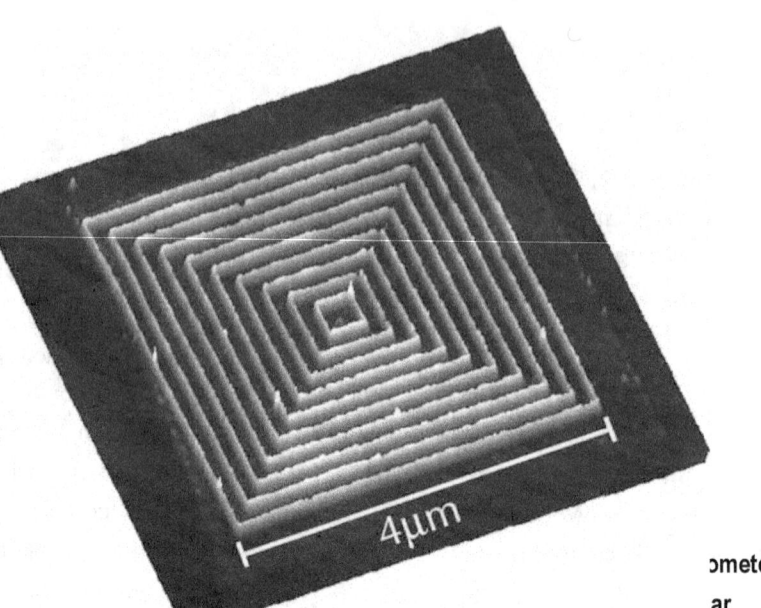

[Using the nan]ometer-[scale Molecul]ar [Measuring Machine], accurate calibration patterns can be produced. For this artifact, the writing method was scanned probe oxidation of hydrogen terminated silicon.

New Photomask Standard, Calibrated in a Unique Industry/NIST Collaboration

The new SRM 5001, the two-dimensional grid standards, were delivered to the NIST SRM office. This represents the first time this new standard 6 inch (15.24 cm) photomask has been delivered to the SRM office. This SRM is expected to provide a traceable standard for the calibration of photomask positioning metrology tools as well as tools that require accurate placement of a wafer within the field of view.

NIST Paper Receives SPIE Metrology Best Paper of 2003 Award

The paper, "Simulation Study of Repeatability and Bias in the CD-SEM," by John Villarrubia, András Vladár, and Michael Postek was selected by vote of conference officials as the best of the approximately 129 papers presented at the 2003 SPIE Conference on Metrology, Inspection, and Process Control.

Invention Disclosures Submitted On Two Revolutionary Advances In Optical Metrology

Researchers from the overlay metrology project submitted two Invention Disclosure and Rights Questionnaire forms (CD 240) for the disclosure of recent potential breakthrough material in high resolution optical metrology. As a part of this disclosure, a new target design that occupies only a few square microns of space was unveiled. The second CD 240 focused on a new proposed method for critical dimension metrology using through-focus focusmetric signatures that have shown near to nanometer-scale sensitivity to changes in linewidth.

Overlay Metrology Methods To Be Adopted By Industry

The overlay metrology project has made available recent important research results on optical characterization, Charge-Coupled Device (CCD) data acquisition calibration, and focus and edge detection work. In-depth discussion were held this year between the NIST overlay project leader and technical representatives from all four of the leading overlay tool manufacturers including Hitachi, Nikon, KLA-Tencor and Sclumberger.

MEMS-scale One Degree Of Freedom High Precision Micro/Nano Positioners Fabricated

NIST designed and fabricated micro-scale versions (i.e., MEMS) of the NIST/MEL one degree of freedom high precision micro/nano positioner, with several feature and dimensional variations. Fifteen silicon dies of this and other calibration devices were fabricated with the collaboration of researchers from the Rensselaer Polytechnic Institute Center for Automation Technologies (RPI/CAT). Each die has a dimension of 10 mm by 10 mm and contains 24 separate devices. The dies were divided among the NIST and RPI/CAT research groups and are currently undergoing performance testing, modeling and calibration.

CRADA with RPI/CAT

A three year cooperative research and development (R&D) agreement between the Center for Automation Technologies, Rensselaer Polytechnic Institute and NIST was signed. The two organizations will be working on a project titled: Modeling and Performance Study of MEMS Positioning Devices.

Field Ion Microscope Images Of Nanotubes Obtained In Collaboration With George Washington University

In a collaboration established in early 2004 between the George Washington University and researchers in the NIST Atom-based Dimensional Metrology project, significant progress has been made in characterizing the ends of the nanotubes by directly imaging the nanotube emitters.

Application of Model-Based Library Metrology to Resist Lines

A study of the model-based library (MBL) technique for SEM critical dimension metrology applied to resist samples has been completed. The NIST-developed technique was applied to polycrystaline silicon samples in previous years. This study of its application to resist samples was motivated by the importance of resist measurements to process control in semiconductor electronics manufacturing.

Additional Accomplishments

- Guest Researcher Hui Zhou completes his Final Defense of his Doctoral research
- "Integrated Metrology: Effective Hardware and Control Strategies" panel chaired by NIST
- "High-resolution Optical Overlay Metrology" presented SPIE Microlithography symposium in Santa Clara
- New Silicon step flow model developed at NIST
- Brad Damazo presentation on the new diode laser interferometry system
- Invited presentation at the ASME International Conference - "Nanotechnology: Measurements and Standards for Manufacturing"
- Invited paper at the Nanotech 2004 Conference
- Reflection mode optical measurements of phase shifting photomasks completed
- New Type MIT Micro Positioner tested
- NNI Research Directions II Workshop participation and invited presentation summarize the Grand Challenge Workshop on Instrumentation and Metrology by Michael Postek
- Invited Presentation "Nanometrology a Fundamental Need for Nanotechnology" at the Microscopy, Metrology and Manipulations using Electrons, Ions, and Photons for Nanophase Materials Workshop
- NIST participation at the NNI Strategic Planning Workshop
- Nova Measuring Instruments sends high level corporate staff for optics discussions
- NIST and SEMATECH collaborate to develop wafers to test new high resolution optical CD techniques
- Nick Dagalakis has been asked to serve as a liaison between the Robotic Industries Association (RIA) office of standards development and the, newly established ANSI/NSP
- Nick Dagalakis of ISD and John Kramar of PED have prepared a CRADA application with APNanotech, Inc.
- Transmission and Reflection Electromagnetic Scattering Theory Model Developed
- Picometer Interferometer Design Described to ASPE
- NIST Assists ISMT in Benchmarking Scatterometry Equipment
- Nanotech in Microlithography Technical Group Meeting held at SPIE 2004
- Successful SEM/XRM Course at NWAFS Spring Meeting
- Workshop on Electron Beam/Specimen Interaction Modeling at SCANNING Meeting
- Invited talk at NSF Workshop. B. Damazo gave an invited talk at the NSF workshop on micro-manufacturing
- Demonstrated a novel method for performing nano-imprint lithography on refractory metal substrates
- Developed a model for describing scanning probe oxidation kinetics of arbitrary materials systems using fractional reaction-diffusion equations
- Demonstration of the principal of Image Stitching Linewidth measurement and publication of two papers on the subject
- AFM measurements of Single Crystal Critical Dimension Reference materials (SCCDRM) specimens for EEEL and SEMATECH members
- Comparison of AFM-based line edge roughness measurements using SEMATECH prototype standards

FY2005 Projects

(Project descriptions can be found within the associated objective)

- Scanning Electron Microscope (SEM) for Nanoscale Measurements (Objective #1)
- Optical Metrology (OM) for Nanoscale Measurements (Objective #2)
- Atom Based Metrology for Nanoscale Measurements and Standards (Objective #3)
- Scanning Probe Microscopy for Nanoscale Measurements (Objective #4)
- Force Metrology for Nanoscale Measurements and Standards (Objective #5)
- Advanced Control Systems and Positioning for Nanoscale Measurements and Standards (Objective #6)
- Optical Tweezers for Nanoscale Manipulation and Metrology (Objective #7)
- Advanced Lithography for Nanoscale Measurements and Standards (Objective #8)
- Development of Nanomachining Technologies for Nanomanufacturing (Objective #9)

Typical Customers and Collaborators

- University of Maryland
- Advanced Micro Devices
- Soluris
- International SEMATECH
- Hitachi High Technologies
- FEI Company
- E. Fjeld Company
- Spectel Research
- INTEL
- KLA Tencor
- Semiconductor Research Corp.
- Photronics and Dupont Photomask
- Nova Metrology tools
- George Washington University
- NASA
- Johns Hopkins University – Applied Physics Laboratory
- Center for Automation Technologies, Rensselaer Polytechnic Institute
- Zyvex Corporation
- Newport Corporation
- Luna Technologies, Luna Innovations
- Duke Scientific
- University of Akron
- Georgetown University Medical Center

FY2005 Standards Participation

SEM Metrology
SEM magnification and linewidth standards are being fabricated; ASTM E42.14, ITRS metrology Technical Working Group (TWG).

Optical Metrology
Extensive participation in efforts at ISMT in the development of benchmarking efforts and standard measurement practices; representation on the Semiconductor Equipment and Materials International (SEMI) standards committees including co-chair for the Microlithography committee; development of photomask 2-dimensional grid standards as well as photomask standards and measurement practices within SEMI.

Atomic scale
Broad participation in efforts at ISMT in the develop of test structures and in the advanced semiconductor needs for linewidth metrology; representation on the SEMI standards committees including co-chair for the Microlithography committee.

SPM
Committee participant - ASTM E42.14 on STM/AFM; ASME B46 on the Classification and Designation of Surface Qualities.

Force
Demonstrated capability of providing SI (International System of units) traceable force as called for by ISO 14577-1,2, and 3 (Metallic materials — Instrumented indentation test for hardness and materials parameters).

Control
Participation in the Optoelectronics Assembly Subcommittee of the IPC, several standards for the handling, attachment, alignment and testing of optoelectronics are currently being drafted.

FY 2005 Measurement Services

Calibrations
- Linescale Interferometer calibration of length scales

Special Tests
- 2D grids and Overlay Measurements (internal)

SRMs
- SRM 2800 – New Optical Microscope micrometer
- SRM 2059 – New Photomask Linewidth (in process)
- SRM 2120/RM 8120 – SEM Linewidth (in process)
- SRM 2090/RM 2090 – SEM magnification
- SRM 2091/RM 8091 – SEM Sharpness
- SRM 5000 – Optical Overlay
- SRM 5001 – 2D Grids

Programs of the Manufacturing Engineering Laboratory

Nanomanufacturing

	2004	2005	2006	2007	2008	2009
Themes	Developing the nanometrology infrastructure for nanomanufacturing					
Principal Activities	Improve imaging and metrology (SEM, AFM, optics)					
	Extend nanofabrication (E-beam, atom-based, imprint, nanomachining)					
	Facilitate control and assembly (high precision stages, optical tweezers)					
Deliverables	New metrology SEM		Extended SEM model	Validated SEM modeling		Reference SEM
	Optical hardware and modeling for 65 nm node			Extension to 32 nm node	Scatterfield microscopy	
	New Litho. STM	Improved etching	Improved tip	Evaluated nanotube tip	3 nm test feature	
	Article on XSM	Traceable AFM artifact	Stitching technique	Techniques Comparison	AFM Meas. Service	
	Article on nano-force		10 piconewton resolution instrument		Intrinsic force standard	
	2x2 MEMS-based positioning stage		6 degree of freedom fiber optic sensor			
	Nanodevice with nanowire		Particle count standard		OT Instrument with 3D visualization	
	SPM Oxidation mask	1&2D calibration structures		Sub-50 nm structure via imprint	Standard template	
	Machining survey	Metrology Frame pub.	Micro/nano imprint pub.		Nano-machined nanoimprint m.	
Impacts	New measurement capabilities	Standards, reference materials	Improved equipment performance measures	Improved modeling	Improved SPM lithography	Calibration services, standards

Programs of the Manufacturing Engineering Laboratory

Smart Machining Systems

Goal

Develop, validate, and demonstrate the metrology, standards, and infrastructural tools that enable U.S. industry to characterize, monitor, and improve the accuracy, reliability and productivity of machining operations, leading to the realization of autonomous smart machining systems.

Program Manager:
Alkan Donmez

Total FTEs:
9

Annual Program Funds:
$2.757 M

Customer Need & Intended Impact

Machining systems are used for discrete part and tooling fabrication and hence are integral to the manufacture of durable goods. Annual U.S. expenses on machining operations total more than $200 billion, about 2 % of Gross Domestic Product (GDP). A study conducted by the Association for Manufacturing Technology (AMT) in 2000 indicated that the advancements in machine tools and related manufacturing technologies created benefits worth a total of nearly $1 trillion in the U.S. over the period 1994-1999. These benefits resulted from gains in productivity, declines in inventory requirements, and manufacturing related product improvements for price, quality and energy efficiency.

Over the last two years there has been an intensive effort originated by three NIST/MEL programs, Smart Machine Tools, Predictive Process Engineering and Intelligent Open Architecture Control that have evolved toward a common theme of Smart Machining. These MEL efforts joined with the National Science Foundation (NSF) and Integrated Manufacturing Technology Initiative (IMTI) to organize and conduct a Smart Machine Tool Workshop in December 2002 bringing government, industry and academia together to identify the U.S. needs in technology development in the area of smart machine tools and machining systems. As a result of this workshop, U.S. manufacturing industry represented by Association for Manufacturing Technology (AMT), National Center for Defense Manufacturing and Machining (NCDMM), National Center for Manufacturing Sciences (NCMS), National Coalition for Advanced Manufacturing (NACFAM), National Tooling & Machining Association (NTMA), Society of Manufacturing Engineers (SME), and TechSolve, established the Coalition on Manufacturing Technology Infrastructure (CMTI) in 2003. This coalition produced a technology roadmap for the Smart Machine Platform Initiative (SMPI) in March 2004. The original three MEL programs strongly influenced the structure and content of the SMPI technology roadmap. The Smart Machining Systems (SMS) program continues this evolution and is closely aligned with the SMPI technology plan.

CMTI indicates in its 2004 Technology Plan that productivity and quality gains achieved by the U.S. manufacturing industry over the last decade are challenged by low-wage countries. As a result, outsourcing of manufacturing in economically critical industries such as automotive, aerospace, consumer products, and heavy equipment is increasing. On the other hand, advanced technologies and engineering innovations are bred in advanced manufacturing environments facilitated by significant amount of interactions. Losing these manufacturing environments, the U.S. is in danger of losing its edge in advanced technology and innovations as well.

CMTI identified an urgent need to reverse this trend by "reinvention of the basic manufacturing environment, enabling dramatic improvements in the productivity and cost of designing, planning, producing, and delivering high-quality products within short cycle times." CMTI further identified five primary thrust areas to address the challenges facing the U.S. manufacturing sector that produces metal parts and fabrications:

a. Process definition and design

b. Smart equipment operation and process control

c Fundamental process and equipment understanding

d. Health monitoring and assurance

e. Integration framework

Metrology and standards are identified as key enablers of these thrust areas. The Smart Machining Systems (SMS) program aims to facilitate the development and validation of such measurement and related technologies and standards.

A successful program will enable cost effective manufacture of first and every part to specification and on schedule by the smart machining systems. Such systems will complement and enhance the skills of machine operators, process planners and design engineers in the manufacturing enterprise by sharing the knowledge and information among these functions to optimize the design and manufacturing processes to their fullest. Loaded with high fidelity process and performance models and optimization tools, smart machining systems will behave in a predictable and controllable manner. This will eliminate trial-and-error based prototype development and reduce time to market, and thus advance the capability of U.S. industry to respond to the global pressures of mass customization of high quality products.

An advanced manufacturing environment is conducive to engineering innovations. Reversing the trend of outsourcing to low-wage countries will enable U.S. industry to regain its competitive edge in innovations and productivity. This competitive advantage will minimize the adverse effects of trade imbalances on the U.S. economy.

Programs of the Manufacturing Engineering Laboratory

Technical Approach & Program Objectives

To enable cost effective manufacture of first and every part to specification and on schedule, a smart machining system will have the following characteristics:

- It will know its capabilities/condition and will communicate this information
- It will monitor and optimize its operations autonomously
- It will assess the quality of its work/output
- It will learn and improve itself over time

These characteristics require a science-based understanding and unambiguous representation of material removal processes and machining system performance.

There are three programmatic focus areas:
(1) performance characterization and representation;
(2) process optimization and control; and
(3) condition monitoring.
Development of dynamic and global optimization tools and methodology that will integrate the physical understanding of all system components will be the unifying theme of all these focus areas.

To meet program goals and objectives as well as communicate the applications of developed concepts and tools to stake holders, the program will focus on three types of projects: development of fundamental methods and data; development of demonstration platforms; and high-risk projects with potential paradigm changing outcomes. Demonstration platforms will also serve to promote stronger collaboration with equipment/software vendors leading to better outreach.

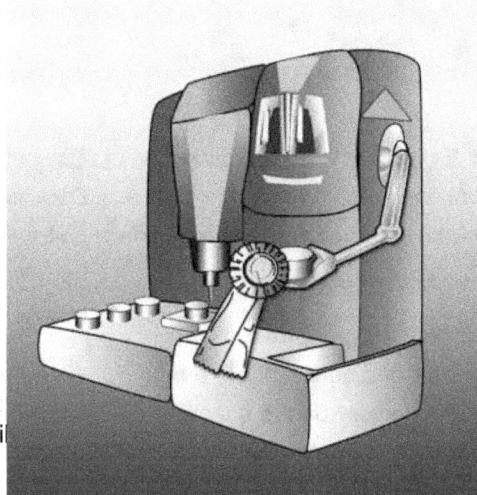

Objective #1: Dynamic optimization

Develop a generic methodology and associated data and modeling specifications to carry out dynamic process optimization, based on design requirements, integrating all related process and equipment knowledge and information.

As stated in SMPI Technology Plan, the ability to account for and accurately predict or describe the propagation of errors in a machining platform is vital for estimating and emulating real-world performance, but represents a major gap in the current technology. Although significant information related to performance of machine tools, machining processes, cutting tools, and materials already exist, there is no unified methodology to combine all this information to generate optimum machining conditions with expected outcomes. Furthermore, very little of this information is standardized, making the optimization even more difficult to generalize.

Accomplishing this objective will enable science-based process design and quality control, which are key requirements for smart machining systems. A generic optimization capability based on well-defined cause and effect relationships will also be an enabler for reasoning and learning capability of smart machining systems.

Programs of the Manufacturing Engineering Laboratory

Objective # 2: Equipment characterization

Develop measurement methods, models and standards to characterize and communicate the machine tool performance under operating conditions.

Information about machine tool performance forms one of the primary foundations necessary to enable manufacturing the first and every part to specifications. Traditionally machine tool performance is determined using a series of tests conducted under quasi-static conditions. There are series of national and international standards describing these tests. These performance parameters are used to buy and sell machines as well as to predict the capability of machine tools for specific family of parts.
The differences between the national and international standards cause the vendors and the users of the machines tools great difficulty and confusion about the claimed performance parameters for contractual and capability estimation purposes. Harmonization among these standards is considered a first priority for this objective. Furthermore, the relationships between the quasi-static performance parameters and obtainable part tolerances are not very well defined because under operating conditions the performance of the machine is not the same as for the quasi-static conditions. The AMT roadmap targets an 80 % improvement in accuracy of machine tool between 1995 and 2010. Machine tool vendors and users have already exhausted their options to improve the performance based on quasi-static machine behavior. Measuring and modeling of performance under operating conditions are the main enablers left to improve machine performance.

Objective #3: Next generation NC

Develop, implement and demonstrate all necessary STEP-NC compliant interfaces and data specifications for seamless operation of model-based machine control.

Smart machining systems need a rich set of information to fully exploit their capabilities. Current Numerical Control (NC) programs are written in "G codes" which express primitive tool paths. These programs do not include information about as-is or to-be geometry, features, tolerances, material properties, fixture location, material removal rates or other information developed during the design and process planning stages. This information is stripped out when converting to G codes, severely limiting the ability of the controller to optimize machining or react to disturbances. Fine tuning processes to maximize performance with current methods is very expensive, tedious and time consuming, and cost effective only for very large part lots. Mass customization and penetration to small manufacturers remain elusive. STEP-NC, an international standard - ISO 14649 "Data model for computerized Numerical Controllers," is the enabling standard that provides the potential for using the digital product model as machine tool input. STEP-NC extends STEP (ISO 10303) – the STandard for the Exchange of Product model data into the NC world.

Boeing and Fanuc recently carried out a case study to demonstrate the impact of incorporating more information into NC process. In this study, Fanuc augmented their G code language with codes signifying the tolerances associated with subsequent tool motions. Boeing's process planners used these G codes to indicate where roughing passes and final finishing passes were taking place. Fanuc's Computerized Numerical Controller (CNC) adjusted its dynamic parameters, such as speed through corners and the amount of blending allowed, and demonstrated a 30 % reduction in machining time. This simple example underscores the drastic improvement that can be achieved in smart machining, when proper information is utilized by the controller.

Information/data generated by process and equipment models and effectively utilized by the controller will enable parts to be machined more quickly with fewer dry runs, prototypes and scrap; improved surface finish and material properties such as work hardening and residual stress; and less variation in parts over large lots. These benefits will accrue to large-volume manufacturers, who can achieve a higher percentage of time-in-cut; and also to small-volume manufacturers, who can save time and cost on prove-outs.

Objective #4: Condition monitoring and reliability

Develop and validate measurement, sensing and analysis methods and associated data specifications and metrics to verify that machine and the process are operating within expected design limits.

Unscheduled downtime of manufacturing equipment is one of the most important impediments to achieving cost-effective, timely production schedules. A recent study by the maintenance and Reliability Center of the University of Tennessee revealed that current estimated industrial maintenance expenditure in the U.S. is estimated to be $500 billion to $700 billion. The same study estimated that development and implementation of condition-based and pro-active maintenance technologies could save $100 billion to $200 billion annually. One example for such high cost of maintenance is that in a large aerospace manufacturing typical life for a machine tool spindle averages between 40 and 400 hours of operation and the plant consumes 1000 spindles per year. If one considers the cost of stopping the production and replacing the large spindle unit at this frequency, even without a damaged tool or the workpiece, it is apparent that the cost savings will be enormous with a smart machining system that monitors its own condition.

Programs of the Manufacturing Engineering Laboratory

Several industrial partners are very interested in collaborating with us in the area of condition monitoring of spindles. Timken, producer of spindle bearings, and Ford Motor Company, a major spindle user, are providing equipment to help us to do research in this area. The spindle test stand that Ford is considering loaning to us is a very unique facility to emulate the real life operation of variety of spindle types used in machine tools. With this test stand we will be able to carry out controlled experiments to correlate the deterioration of spindle performance and spindle condition data. The use of such a test stand allows tuning and improvement of algorithms and sensor applications for spindle condition monitoring and condition based maintenance. Such a system will avoid unscheduled downtime and enable cost savings mentioned above

Objective #5: In-situ metrology

Achieve a breakthrough in advanced metrology to enable direct measurement of the pose of the cutting tool or measuring probe relative to workpiece with an uncertainty better than 5 m and 2 arcseconds in a 1 m^3 cubic workzone.

Most accuracy problems in machine tools and Coordinate Measuring Machines (CMMs) are related to indirect measurement of the tool or probe positions with respect to the workpiece. Due to large Abbe offsets introduced by stacked slides configurations of traditional machine tools these indirect measurements cause large uncertainties in the measurements.

Because of these uncertainties and the uncertainties introduced by the machining process, machined parts have to be inspected to verify that they are within the design specifications. Post process inspection is very costly non-value added activity requiring CMMs or other custom designed inspection systems. Eliminating this activity would save significant time, money and other scarce resources such as skilled labor. We know that Boeing and Caterpillar are very interested in implementing independent measurement systems in their machine tools. Although it is a high risk effort, achieving this objective will, at a minimum, enable direct on-machine measurements and certification of machined parts. Furthermore, a successful system would allow for breakthrough improvements in machine accuracy. It dramatically lowers current requirements and costs for accurate slideways, carefully controlled heat sources, a thermally invariant and stiff machine tool structure, and a solid foundation. Finally, such a system will significantly lower the requirements for stable environmental conditions, one of the highest hidden costs to achieving precision.

Major Accomplishments

Calibrating Kolsky Bar Temperature Measurements with Pure Aluminum and Zinc Samples

Performed experiments where pure aluminum and pure zinc samples were melted to aid in the calibration of the pyrometers and thermal imaging cameras used in Kolsky bar apparatus. The melting point experiments provide two known temperatures to record and confirm the calibration of the instruments.

Cutting Temperature and Force Measurements of Aluminum and Titanium Alloys Performed

Measured machining temperatures and forces for orthogonal cutting of titanium (Ti6A14V) and aluminum (AL7075-T651) using dual-spectrum, high-speed video and three-axis force measurement. Preliminary analysis of the results indicates that the lower thermal conductivity of the ceramic tool material appears to have the predicted effect of directing more of the heat from the cutting process into the chips. Comparisons between measurements and model-based predictions indicate generally favorable agreement for cutting forces, with opportunities for improvement in prediction of thrust forces through improved friction modeling.

Development and Deployment of New Optical Configuration of High-Speed Infrared Camera

A new optical configuration for the high-speed infrared video camera involving a new reflective lens with a larger depth of field was developed. This new configuration yields a significant improvement in depth of field over the prior configuration, yielding sharper images of target objects with rough surfaces allowing us to image the rough side the chip at the tool-workpiece interface.

National and International Standards for Machine Tool Performance

We continued to provide the Secretariat for ISO/TC39/SC2 "Test Conditions for Machine Tools" and for the American Society of Mechanical Engineers (ASME) B5/TC52 "Performance Evaluation of Machine Tools." Accomplishments include: 1) New revision of ASME B5.54 "Methods for Performance Evaluation of Computer Numerically Controlled Machining Centers;" 2) Contributions for major restructuring, modernization, and harmonization of ISO 230-1 "Geometric Accuracy;" 3) Publication of ISO 230-9 addressing measurement uncertainty of obtained performance parameters; 4) Development and extensive experimental evaluation of test methods and performance parameters for errors due to the heat generated by high-speed axis motions for the revision of ISO 230-3; 5) Conducted tests for evaluation of environmental vibration effects on machine tool structure and provided results for inclusion in ISO/CD 230-8 machine tools test code for determination of vibration values; 6) Experimental study on the application of a laser vibrometer to test error motions of ultra high-speed spindles; 7) Collaborative report to ISO and ASME on test procedures for the alignment of machine axes, with a focus on four- and five-axis machining centers; and 8) Eight ISO Draft International Standards (DIS) and three Committee Draft standards (CD) on machine tool test methods.

Machine Tool Data Standards

We continued to provide the Secretariat for the ASME B5/TC56 Standards Committee "Information Technology for Machine Tools." Accomplishments include: 1) Ballot ready draft standards ASME B5.59-1 (performance test ___) ASME B5.59-2 (machine t___ properties); 2) Reports of t___ activity to ISO TC39 (machine tools) and TC18___ (Industrial automation systems and integration) gaining support for an ISO activity; 3) Updated reference schemas, example files, and style sheets, to facilitate implem___ tation, field-testing, and de___ stration of the standards; a___ 4) Demonstration of self-kn___ the turning center testbed through a description of its properties and performance data according to the information model in B5.59-1 and B5.59-2.

Smart Transducer Standards

We still lead the development of suite of standards associated with smart transducer applications. Accomplishments this year include: 1) The Institute of Electrical and Electronics Engineers (IEEE) 1451.4 standard "Smart Transducer Interface for Mixed-Mode Communication Protocol" was re-balloted and approved as a standard on May 2004. It will be published by IEEE in 2004.; 2) Developed the IEEE 1451.1 neutral model (data and object models) in JAVA and published them in the IEEE 1451 open-source repository.

Prediction of Workpiece Errors

Developed prototype capability to translate machine tool performance parameters into error bounds for the form and dimensional errors of elementary machined features. The developed techniques ___ an error-budget approach ___ n the laws of uncertainty ___ agation to detailed Monte ___ arlo simulations, and take ___ nto account the effects of ___ uncertainties in machine ___ errors. Measurements were ___ performed on a machining ___ center to assess and model ___ key geometric and thermal ___ rror sources. The results ___ re used to predict the error ___ ds of features on a series of ___ chined at different times.

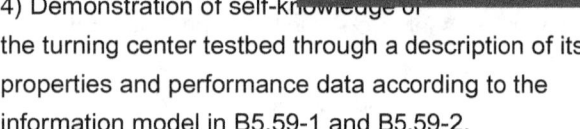

STEP-NC Performance Testing

Developed interpreters for two components of STEP-NC, the underlying data model (ISO 14649) and the STEP-integrated model (ISO 10303 AP-238) to investigate the computational burden for parsing data files in real time. Tests indicated that the performance of the interpreters was comparable, and that no significant penalty executing the STEP-integrated model was incurred.

Programs of the Manufacturing Engineering Laboratory

STEP-NC Conformance Testing

In collaboration with The Boeing Co., test procedures were designed to determine if STEP-NC data for complex curved parts can be executed the same way on NC machine tools from various manufacturers. The data includes materials, cutters and tolerances as well as traditional geometry and features. NIST's 5-axis machining center with a Siemens 840D CNC complements several Boeing machines in both machine configuration and NC control. The tests uncovered numerous discrepancies between machines, and Boeing is implementing short-term workarounds while the underlying problems are being resolved.

STEP-NC Data Visualization Software Written

The software to visualize tool paths resulting from the execution of STEP-NC (ISO 14649) machine tool programs was developed. The new visualizer can be used to draw tool paths for the other supported programming languages, making it more broadly useful.

FY2005 Projects

Robust optimization of machining (Objective #1)

A mathematical framework will be developed to define dynamically adjusted objective and constraint functions for optimizing the whole machining operation taking into account machine performance, process capability, part design specifications, time and cost.

Physics-based modeling of machining (Objective #1)

A dynamic property model (constitutive model) for the material to be used in the optimizer demonstration case for the first project will be developed. The project will continue experimental work in (1) orthogonal machining with force, thermal and visible video measurements of tool and chip interface, and (2) dynamic material property measurements with pulse heated Kolsky bar apparatus.

Machine tool performance characterization and improvement (Objectives #2 and #5)

This project will focus on the following major activities: (1) national and international standards development, (2) development and field testing of standardized information models and associated data formats for machine tools, (3) implementation of Bayesian approach to machine tool performance modeling and quality control, (4) predictive tolerance analysis of machined parts, (5) exploration of advanced metrology enabling machine self calibration, in-situ part inspection, in-situ dynamic tool characterization, and improved feedback on realized pose of the cutting tool.

Next generation NC for smart machining systems (Objective #3)

Data and interface requirements for adaptive control implementations to extend the capabilities of existing NC standards will be identified and documented. There will be two parallel activities: (1) review STEP-NC data model and its support for adaptive control, (2) install Siemens STEP-NC software onto the 5-axis machining center controller to carry out initial testing of its capabilities. Tests will include Roy-G-Biv's Extensible Markup Language (XML) based controller interface, XMC, for model-based control data exchange.

Model-based machine tool control (Objectives #2 and #3)

This project will be the demonstration platform for the development of techniques to improve the efficiency of turning operations at Picatinny Arsenal and its supplier base through the application of open-architecture model-based machine control to both new and existing machine tools.

Smart spindles: Testbed verification (Objective #4)

The requirements for real-time spindle condition monitoring and diagnosis system will be investigated. This will be based on sensing and signal processing scheme, which involves preliminary optimization of sensor placement, under static and dynamic radial loading conditions.

Typical Customers and Collaborators

Industry:
- Alibre
- Alcoa
- AMT
- API
- Boeing
- Bosch-Rexroth
- Caterpillar
- Cincinnati Lamb
- DaimlerChrysler
- EDS PLM
- Esprit
- Ford Motor Company
- General Dynamics
- GE Fanuc
- General Motors
- Gibbs & Assoc
- Hardinge Brothers
- Heidenhain
- IBM
- IMTI
- IQL
- Knowledge Based Systems
- Lion Precision
- Max Spindle
- MIMOSA

- Northrop Grumman
- Okuma
- Optodyne
- Pratt & Whitney
- Renishaw
- Roy-G-Biv
- Siemens
- STEP Tools Inc
- Third Wave Systems
- Unova
- VulcanCraft

Government:
- Army Picatinny Arsenal
- Los Alamos National Labs
- NASA
- Oak Ridge National Labs
- Sandia Labs (potential)

Universities:
- University of Florida
- University of Maryland
- University of North Carolina at Charlotte
- Southern University
- University of Massachusetts
- University of Aachen, Germany
- Pohang Institute of Science and Technology, Korea
- University of Loughborough, UK
- University of Auckland, New Zealand FY2005

Standards Participation

- ANSI/ASME B5 Machine Tools (Member)
- ANSI/ASME B5/TC52 Machining & Turning Centers (Secretariat)
- ANSI/ASME B5/TC56 Information Technology for Machine Tools (Chair and Secretariat)
- ANSI/ASME B89.3.4 Axes of Rotation (Observer)
- ISO TC 39 Machine Tools (Member of the US Delegation)
- ISO TC 39/SC2 Test Conditions for Machine Tools (Secretariat)
  - ISO TC 39/SC2/WG1 Geometric Accuracy of machine tools (Member)
  - ISO TC 39/SC2/WG3 Test Conditions for Machining Centers (Member)
  - ISO TC 39/SC2/WG6 Thermal Effects on Machine Tools (Convener)
  - ISO TC39/SC2/WG7 Reliability, capability and availability of machine tools (Member)
  - ISO TC39/SC2/WG8 Vibration of machine tools (Member)
- ISO TC 184/SC 1/WG 7 Data modeling for integration of physical devices (Member)
- ISO TC 184/SC 1/WG 8 Distributed installation in industrial applications (Member)
- ISO TC 184/SC4 Industrial Data (Member of the US Delegation)
- ISO TC 108/SC 5/WG 6 Data Processing and Analysis Procedures for Condition Monitoring and Diagnostics of machines, Including Formats and Methods for Communicating, Presenting and Displaying Relevant Information and Data (seeking active participation)

Programs of the Manufacturing Engineering Laboratory

Smart Machining Systems

First Part Correct   First and Every Part Correct

Downtime, CBM

Special activities

Special activities

MEL is more than the sum of its programs and projects. Our staff works on issues that concern the U.S. manufacturing community by developing new measurements and traceable metrology methods to help U.S. manufacturers compete with their foreign competitors, participating in standards setting activities, and improving manufacturing operations, enhancing international competitiveness, and enabling technology breakthroughs. Most of our work fits in one of our eight strategic programs; however, some activities take place outside the scope of these programs. This section describes these special activities, which take one of many forms: exploratory projects, competence projects, leading the U.S. efforts to participate in international manufacturing activities, supporting a government-wide effort to leverage government funded R&D programs in manufacturing, and taking the leading in a NIST-wide program to support manufacturing through information technology efforts.

Each year the NIST Director sets aside $10 to $12 million for a Competence Program that funds several projects that represent the best ideas of the many innovative, high-risk ideas of NIST's scientists and engineers. These projects often span the traditional activities of the individual NIST laboratories. Before they are funded, the proposals undergo a rigorous peer-review by both internal and external technical reviewers – reviewers from a variety of our Nation's technological industries. Funded proposals display a high level of risk, innovation, and technical quality, a clear research-oriented strategic plan with goals, and a demonstrated and important tie to NIST's mission and its 2010 strategic plan to build new competence at NIST. In 2004, the NIST Microforce Realization and Measurement Competence Project, lead by MEL's Jon Pratt, concluded. A brief summary of the accomplishments by Jon's team appears on page 126. In 2005, a new competence project titled "Phase-sensitive Scatterfield Optical Imaging for Sub-10 nm Dimensional Metrology," led by MEL's Richard Silver, will begin. A description of the project included in this section will be found on page 131.

In 1998, the MEL Management Council introduced a mechanism to examine new areas of research – the MEL Exploratory Project. Each year, a small portion of laboratory funding is allocated for these projects to test the feasibility of a new idea or technology. Exploratory projects focus on topics that are within the mission of MEL, but outside the scope of any current individual MEL or NIST strategic program. Exploratory Projects, typically lasting one year, experience varied outcomes. They could be deemed a success and brought to a natural conclusion at the end of the first year, they could be considered for expansion into a future strategic program, or they could be incorporated into an existing program. A brief summary of both newly-awarded and recently completed Exploratory Projects can be found starting on page 134.

From 2001 until 2004, six federal agencies involved in manufacturing research and development joined together to improve the exchange of information on their technical programs and collaborate to enhance the payoffs from federal investments in manufacturing. This group, the Government Agencies Technology Exchange in Manufacturing (GATE-M), initially focused their efforts on two topics of joint interest: intelligence

Special Activities

in manufacturing, and nanoscale and microscale systems and technologies. To foster information exchange, GATE-M participants conducted detailed interagency reviews of programs. MEL served as the lead agency working with the other agencies planning and coordinating the group's activities. In FY2004, MEL led the transition from GATE-M to a new Interagency Working Group (IWG) on Manufacturing Research and Development (R&D). The IWG operates under the auspices of the National Science and Technology Council and is chartered by the White House Office of Science and Technology Policy (OSTP). The current group, chaired by the Department of Commerce and vice-chaired by the director of MEL includes at least 14 federal agencies. More details on this group can be found on page 159.

NIST is committed to working with partners in the international community to improve manufacturing operations and enhance international competitiveness and technology breakthroughs through the Intelligent Manufacturing Systems (IMS) Program. IMS participants from Australia, Canada, the European Union and Norway, Japan, Korea, Switzerland, and the U.S. cooperate in research and development projects, share costs, risks, and expertise to advance the next generation of advanced manufacturing and processing technologies. The U.S. Secretariat that facilitates the participation of U.S. organizations is located in MEL. A description of the IMS program can be found starting on page 162.

The Systems Integration for Manufacturing Applications (SIMA) program is an intramural effort underway at NIST to support the application of information technologies to the manufacturing domain. Initiated in 1994, program members work with industry to develop technology solutions that enable integration of the systems used in the engineering and manufacturing of various kinds of products. The SIMA Program Office, managed from within MEL, oversees research projects conducted throughout NIST. These activities improve manufacturing productivity on the shop floor and impact research conducted to devise new products and processes, the design of products and processes, the engineering analysis of prospective solutions, the planning of manufacturing operations, the scheduling of production operations, the engineering of production capabilities, and the myriad of other information-intensive activities performed in industry today. A brief description of this program can be found on page 166.

Competence Projects

Contact:
Jon R. Pratt
jon.pratt@nist.gov

Collaborators:
John A. Kramar, MEL,
David B. Newell, EEEL,
Douglas T. Smith, MSEL

Microforce Realization and Measurement Competence Project

Introduction:

The Microforce Realization and Measurement Competence is proud to report the creation of the NIST Small Force Metrology Laboratory (SFML), the first and at present only laboratory in the world capable of providing accurate force standards to researchers and manufacturers that employ micro and nano-mechanical tests for product development, fabrication, and quality control. Forces from millinewtons down to nanonewtons, commonly measured in industry and academe using devices such as Instrumented Indentation Machines (IIMs) and Atomic Force Microscopes (AFMs), can now be realized in the SFML with traceable relative uncertainties less than 0.01 % at loads near a micronewton. NIST's ability to measure these forces in a fashion traceable to the International System of Units (SI) removes a significant impediment to the creation of quantitative commerce from qualitative science.

The following briefly summarizes the achievements of the Microforce Realization and Measurement Competence team, highlighting the one-of-a-kind Electrostatic Force Balance — a device that accurately realizes an SI micronewton in terms of voltage, impedance, and length metrology. The report reviews the objectives and the significant outputs and outcomes of the project, the current allocation of resources, the collaboration's past, present, and future, and concludes by forecasting a vital future for the SFML as it supports quantitative nanomechanical testing that spans the physical, chemical, and life sciences.

Special Activities

Goal:
Develop methods and apparatus to realize and measure forces between 10^{-8} and 10^{-2} N setting an international standard in the new area of microforce metrology

Objectives:
The project goal as stated in a presentation made to former NIST Director Ray Kammer was:

In support of this goal, the team's objective has been to derive small force standards from the electrical and length units, rather than mass, length, and time, and to do so in a fashion traceable to both the SI and to quantum invariants, taking our cue from the Electronic Kilogram experiment. Within the first year of the project, the project team chose the specific objective of realizing an electrostatic force of nominally 10^{-5} N with relative uncertainty less than 10^{-4}. They achieved this using the centerpiece of the SFML, the Electrostatic Force Balance.

The Electrostatic Force Balance (EFB) is a mechanical null balance, designed and fabricated at NIST, that is very similar to conventional analytic balances or mass comparators, but employs a null force calculable from the gradient of a field characterized using length, voltage, and capacitance measurements traceable to both SI standards and quantum invariants. The EFB is capable of acting as a primary standard of force below 10^{-5} N, and has freed the group from the difficult task of further subdividing the Kilogram below a milligram, although for forces above 10^{-5} N they continue to use deadweight loads whenever practical.

Outputs:
The competence project has created foundational metrology for micro and nano mechanical testing upon which the future of many nanotechnologies will rest. These are:

- A primary standard of force in the micronewton regime that is reliable at the unprecedented level of parts in 10^4
- Prototype artifact transfer standards capable of transferring the unit of force at the micronewton level with uncertainty below one percent
- Methods and apparatus for the traceable calibration of force in IIM equipment
- Methods and apparatus for the traceable calibration of force in AFM equipment

The foundational metrology is embodied by novel equipment, capabilities, and skills housed in the SFML that combine to create an effective resource for the world's scientists and technologists tackling problems that require the quantitative measurement of small forces. The team has become competent at the realization and measurement of these forces in a way that did not previously exist, as evidenced by the following firsts.

1. For the first time, it is possible to assess the performance of instrumented indentation equipment in an objective fashion:

   R.M. Seugling and J. R. Pratt, "Traceable Force Metrology for Micronewton Level Calibration," Proceedings of the American Society for Precision Engineering, 19th Annual Meeting, Orlando, Fla, 2004.

Special Activities

2. For the first time, it is possible to compare the results of AFM experiments against a NIST calibrated reference:

> J. R. Pratt, D.T. Smith, D.B. Newell, J.A. Kramar, and E. Whitenton, 2004, "Progress towards Systeme International d'Unites traceable force metrology for nanomechanics," Journal of Materials Research,19(1), pp. 366-379.

> G.A. Matei, E.J. Thorsen, J.R. Pratt, D.B. Newell, and N.A. Burnham, "Thermal Method of Cantilever Calibration over a 200 kHz Bandwidth," submitted to Nanotechnology, August, 2004.

3. And, for the first time, it has been demonstrated that an experiment in the spirit of the NIST Electronic Kilogram can yield an effective Electronic Milligram with greater certainty than currently available from direct mass calibration:

> J. R. Pratt, D. Newell, J. A. Kramar, and R. M. Seugling, "Realizing and Disseminating the SI Micronewton with the Next Generation NIST Electrostatic Force Balance," Proceedings of the American Society for Precision Engineering, 19th Annual Meeting, Orlando, Fla, 2004.

These achievements resulted in the following recognition:

- Presidential Early Career Award for Scientists and Engineers(PECASE), 2003; Pratt
- Department of Commerce Silver Medal Award, 2004; Pratt, Kramar, Newell, and Smith
- Invitation to the National Academy of Sciences 2004 Keck Futures Initiative; Pratt

Jon Pratt with the NIST Electrostatic Force Balance

Outcomes

The world's first small force metrology laboratory has become the international benchmark, and has profoundly influenced the proposed National Physical Laboratory (NPL) Low-force Measurement Facility in the United Kingdom (Low-force Measurement Facility Mechanical Design Report, September 2004, NPL), and spawned similar activities at the Korean and Taiwanese standards laboratories.

The SFML provides the foundational metrology for standards activities in instrumented indentation and quantitative AFM measurements of adhesion and provides the first real means to achieve industrial compliance with ISO, ASTM and VAMAS (Versailles Project on Advanced Materials and Standards) standards–standards that were written in anticipation of the NIST SFML and its unique capabilities.

Nanomechanical testing tools are used in every manufacturing sector for the development and selection of materials and material properties. The market for these tools is dominated by US companies, one of which is already using the results of SFML data to improve its manufacturing processes, leveraging our work to convince semiconductor manufacturers that instrumented indentation is in fact quantitative and can be relied on for fabrication and quality control, not just process development.

Special Activities

Resources:

As of fiscal year 2005, the SFML has been incorporated into the MEL Nanomanufacturing program, which supports the project at the level of about 0.6 FTE (full-time equivalents). This programmatic support is leveraged at the division level in the Manufacturing Metrology Division, which supports approximately 0.3 FTE, and the Precision Engineering Division, which supports approximately 0.25 FTE, and at the MEL laboratory level through the award of PECASE funds to the principal investigator, Dr. Pratt, amounting to another 0.3 FTE, for a total of approximately 1.5 FTE, with allowances made for other objects spending. As is often the case, critical staffing must be augmented with National Research Council (NRC) postdoctoral researchers, which the project has successfully recruited.

The other participating laboratories, the Materials Science and Engineering Laboratory (MSEL) and the Electronics and Electrical Engineering Laboratory (EEEL), no longer provide direct support, though their researchers maintain access to the SFML and continue to contribute to the project in an advisory capacity. This is important, since EEEL provides vital metrological support to the EFB, and because MSEL scientists are customers for the small force calibrations facilitated by the SFML. For instance, MSEL is the point of contact for ISO, ASTM, and VAMAS standards activities in instrumented indentation and adhesion that ultimately rely on the SFML for traceability.

Collaborators:

The collaboration between MEL, MSEL, and EEEL was highly effective at both a technical and strategic level. The technical expertise required to create the EFB drew on all the participants' accumulated knowledge of advanced manufacturing, precision engineering, and experimentation, to say nothing of the institutional knowledge of mass, length, voltage, and capacitance metrology that was ably communicated and brought to bear on the problem by the respective laboratories.

From a strategic standpoint, the connection of MSEL to the vast nanomechanics community provided a clear, tangible customer upon which to focus. Now, though the project is firmly rooted in MEL, the SFML continues to be a resource for MSEL's nanomechanics and nanotribology groups, as they work to support various ISO, ASTM, and VAMAS standards activities around the globe.

Special Activities

The future:

Given these successes, it is no surprise that the project will continue, and that it is anticipated to play a significant role in expanding MEL's Nanomanufacturing program beyond the program's core expertise in nano-scale dimensional metrology.

The vision for the project's future is to continue serving a base constituency of IIM and AFM users at NIST, other National Metrology Institutes (NMI's), National Laboratories, original equipment manufacturers (OEM) instrument manufacturers, large industrial research laboratories, and academia by developing and refining our dissemination skills, so that all these potential customers can enjoy traceability to an accepted international standard. This means working diligently to improve our efforts at developing transfer artifacts, an area where our European colleagues at PTB (Physikalisch-Technische Bundesanstalt, Germany) and NPL have experience and may be able to help NIST's developments.

The team believes that as force realization approaches the level of molecular and atomic force (nanonewtons for covalent bonds), it will be possible to realize so-called intrinsic force standards. Thus, they intend to continue pushing their capabilities to measure even smaller forces, with a new EFB under construction and projected to measure a nanonewton with uncertainty at the percent level.

Going even further, if optical forces can be accurately cross checked to the SI (piconewtons), we see a huge potential to support nascent work in bio-nanotechnology and the associated field of mechanochemistry, where researchers examine the energy of reaction in terms of work done using single molecule tests with force loading, typically applied using optical trapping techniques. This field is exploding and may finally lead to the unraveling of the very physics of life, providing insight into the mechanisms of protein folding and DNA (deoxyribonucleic acid) binding.

To bring SI force to this molecular level, measure intrinsic standards, and keep pace with the accelerating field of bio-nanomechanics, however, will require an effort comparable to that already expended on the microforce realization competence, with additional contributions likely needed from other areas of physics and chemistry. Thus, development of this capability will serve as the centerpiece of a new NIST Competence proposal to be submitted in 2005.

Special Activities

Contact:
Rick Silver
richard.silver@nist.gov

Collaborators:
Thomas Germer (PL),
Nien Fan Zhang,
Charles Hagwood (ITL)

Phase-sensitive Scatterfield Optical Imaging for sub-10 nm Dimensional Metrology

Summary:

This competence project will advance optical microscopy to unprecedented levels of performance through theoretical and experimental development of a new technique we call "scatterfield optical imaging." This new method promises to make possible optical measurements of nanometer-sized features using high-throughput, low cost optical methods with the potential for an enormous impact on innovation and quality control in semiconductor manufacturing and nanotechnology as well as providing the measurement basis for new calibration standards well beyond the state-of-the-art.

Concepts and the Research Challenge:

Technical Summary

By developing experimental and theoretical-based analytical techniques, initial simulations indicate it is possible to measure sub-10 nm sized features using new, advanced optical methods. The ability to measure linewidth, feature overlay, and defects in nanoelectronic devices using optical methods presents a solution with enormous impact. It is anticipated that features as small as 5 nm may be measured, a factor of 50 extension beyond the practical Rayleigh limits of conventional visible-light microscopy. The feasibility of the basic approach, which uses structured illumination, high-resolution Charged-Coupled Device (CCD)-array processing, engineered target designs, detailed theoretical modeling of electromagnetic scattering, and phase-sensitive data analysis, has been demonstrated. Significant interest in the proposed technique by elements of the semiconductor industry and evolving nanomanufacturing industry suggests applications of this new measurement science and technology could have a major impact on manufacturing quality, process control, and throughput.

Introduction

Although optics are often thought to lose its effectiveness as a metrology tool beyond the Rayleigh resolution criterion, we have evidence that optics can be used to image and measure features smaller 10 nm in dimension. The resolution obtained by conventional optical microscopy is usually considered to be limited to approximately half of the wavelength of the light used for illumination.

Special Activities

However, another optical metrology method, interferometry, can routinely achieve sub-nanometer resolution using visible light. The key to understanding the difference between these two methods is related to the complexity of the electromagnetic fields (illumination) and the targets used in the measurement. In the case of interferometry, single plane waves, that is, waves having a simple sinusoidal oscillation in space and time, are reflected from planar, mirrored surfaces, with the phase information preserved. For microscopy, the illuminating fields are much more complicated and the samples often have significant three-dimensional character.

Optical imaging and metrology with resolution well beyond the wavelength is a possibility as a result of a new optical configuration and design, increased computation speed, high-resolution CCD array technology, improved signal processing of large arrays, and reliable theoretical scattering models. The ability to "engineer" the illuminating field is key to this advanced measurement technique. It involves low numerical aperture (NA) illumination optics, annular or polar illumination schemes, or high angle to on-axis plane wave illumination. Another key technological advance is the realization that phase information, accessible by stepping through the ideal focus, provides a sensitive, complex optical signature for a given sample and optical configuration. The sample or target design also plays a central role in the ability to increase optical sensitivity to dimensional changes. A target structure which enhances the amplitude of the scattered fields and accessible information content can be optimized with accurate simulation tools.

Nanometer size features may be measured optically by scatterfield optical imaging technique

The new proposed measurement technique has some conceptual ideas parallel to scatterometry and other aspects parallel with interferometry, but adapted to a flexible optical configuration, a unique, new powerful capability exists. The parallel with interferometry is the fundamental use of the phase information. Measuring the intensities through focus (different heights relative to the sample) is sampling the phase information by measuring different slices of the electromagnetic fields. In scatterometry, typically a beam is collimated and the zero or first-order reflected intensity is captured. In this new configuration, the entire scattered field (all accessible orders) is collected and imaged. A major advantage of this methodology is that, unlike scatterometry, having optical imaging components (lenses) allows the image to be magnified, allowing targets of only a few micrometers in size to be measured as opposed to the 50 μm by 50 μm targets measurable by scatterometry. In addition, the optical design allows a broad array of tailored illumination schemes to be used. This becomes a very powerful tool when combined with well-engineered target designs.

Special Activities

A key enabling technology that these improvements rely upon is the unprecedented accuracy by which we can now calculate the electromagnetic scatter from features. The exact profile of the scattered field is extremely sensitive to small changes in the shape and size of the scattering feature. Once one can accurately calculate the scattering function and propagate it in the optical system, one can in principle measure features near to nanometer dimensions. It is no longer necessary to be limited by the concept of image-based edge detection microscopy in optically-based dimensional metrology. This is a radical departure from traditional methods of optical imaging.

Impact

The impact from the successful implementation and development of this new measurement science will be enormous. Optics has some significant advantages over other high resolution imaging techniques such as scanning electron microscopy (SEM) and atomic force microscopy (AFM). SEM requires high vacuum with expensive instrumentation and can be destructive in making measurements. AFM has significant throughput and reliability challenges. Optical metrology, on the other hand, is massively parallel, with high throughput, relatively low instrumentation cost, and is non-destructive. If we demonstrate the ability to accurately measure sub-10 nm features with nm sensitivity to changes in shape, the impact will be significant.

There is a current push in semiconductor manufacturing to use tighter closed-loop process control – this trend will likely be followed in nanomanufacturing applications. This is known as integrated manufacturing and several companies are developing limited hardware for these applications. The approach described here can have a major impact on integrated manufacturing capabilities.

Special Activities

Exploratory Projects

The MEL Management Council annually sets aside funds for exploratory projects. The MEL Management Council (MELMC) wanted to give employees an outlet to explore new technologies, ideas or ideas of research that were outside of the current scope of the MEL Strategic programs. Exploratory projects are also an excellent opportunity for the MEL staff to submit their ideas on new technical areas in which MEL might become more involved in the future. The short duration of the exploratory project (i.e., approximately one year) is sufficient time to test the feasibility of the new idea or technique. Topics are within the mission of MEL, but outside the scope of any current individual MEL or NIST strategic program. Since benefits from these projects should accrue to MEL as a whole, emphasis is also placed on project ideas that cross division or laboratory lines.

In FY2005, the MELMC received 35 pre-proposals that included over 50 different authors, including representation from all five MEL divisions, two other NIST Laboratory Operating Units, and four external collaborators. Goals, accomplishments, and plans for follow-on work are described below for each of the FY2004-FY2005 exploratory projects. Following the current Exploratory Projects, you will find descriptions of the recently concluded exploratory projects.

Special Activities

Contact:
John Dagata
john.dagata@nist.gov

Advanced imaging for nanoparticle drug delivery systems

Nanoparticle drug delivery systems (NDS) are a generic technology that is currently under intense development by the healthcare community for a broad range of applications. The NDS payload may consist of not only gene-carrying DNA but, for instance, regulatory proteins that turn on or off specific cellular processes. The dimensional properties of these systems profoundly impact the efficacy of treatment, e.g., stability of an NDS in the bloodstream and its transfection rate (the ability of an intact NDS to cross a cell membrane). Reliable methods for imaging molecular components and sub-assemblies of these soft materials with resolution well below 100 nm are therefore in demand for quality control during clinical and pre-clinical trials of prospective treatments as well as for fundamental understanding and control of NDS self-assembly.

MEL is presently collaborating with Esther H. Chang, Professor of Oncology, Georgetown University Medical Center (GUMC), Washington DC and Synergene Therapeutics, Inc., Washington DC to demonstrate high-resolution imaging of individual components and associated complexes of the constituents of a NDS (Tf-Lip-p53) and imaging of the transfection process with fixed and live cells. The technical approach employs MEL's extensive knowledge of simultaneous topographic, electric force, and phase imaging capabilities of a Scanning Probe Microscope (SPM) to assign a unique signature to each material, and then to optimize the sensitivity of SPM operation under appropriate preparative and physiological conditions.

This development will be done in parallel with confocal fluorescent imaging of the cells. We will compare optical and SPM information of cell transfection rates with scanning electron microscope (SEM) imaging of fixed cells using recently developed fluid imaging techniques in conjunction with MEL's recently acquired Hitachi S-4800 instrument operating in very high resolution digital imaging mode. This SEM offers a wide range of electron landing energies and sample biasing in reflected and transmission electron detection operation modes that are very advantageous for these experiments.

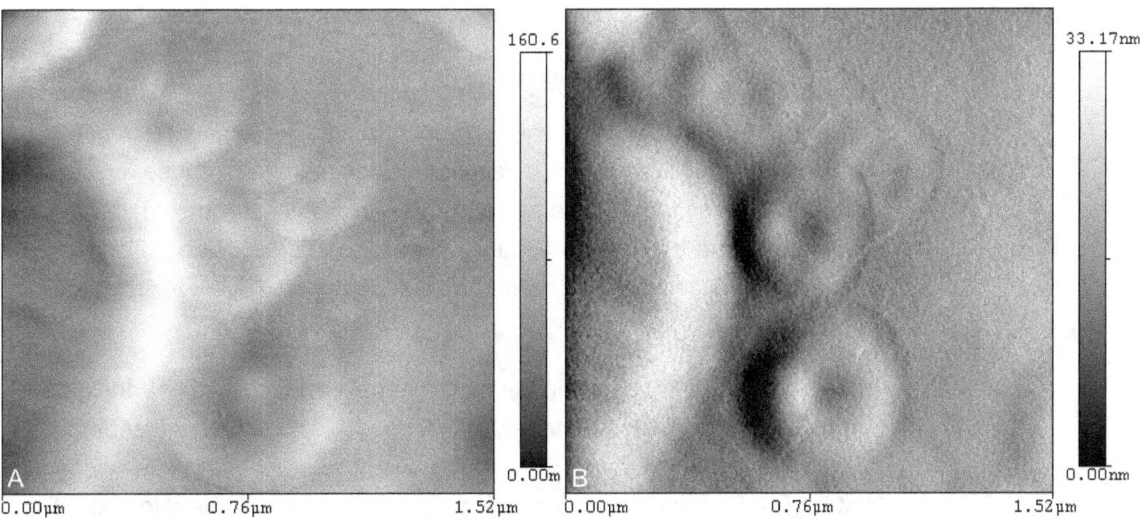

Scanning probe microscopy of ~300-nm diameter liposomes containing MagnaVist, a magnetic resonance imaging contrast agent: (A) magnetic force/phase image, and (B) topographic image. Liposomes prepared at Georgetown University Medical Center and imaged at NIST.

Special Activities

Contact:
Johannes Soons
soons@nist.gov

A Bayesian Approach to Model-Based Quality Control of Machined Parts

The objectives of this project are to develop a new systematic method to capture and improve knowledge on the accuracy of a machine tool and then come up with a proof-of-concept of new method. The resulting model can be applied to adjust the machine or to evaluate its capability to machine a specific part. The method enables the model to learn from its mistakes while addressing the challenges associated with the diverse and often incomplete data on machine errors encountered in industrial environments. The method is a new approach to Statistical Process Control (SPC) to address agile environments. A key part of the project is the application of the method to a milling machine in the MEL's Fabrication Technology Division (FTD), using inspection results of parts machined during normal operations spanning an entire year.

SPC is a technique used by industry to ensure that parts are manufactured to specification. It relies on a statistical analysis of post-process inspection results to identify trends in the quality of the manufactured parts. The results are used to make adjustments to the machine tool(s) or process parameters. SPC is successfully used when making large quantities of parts under similar conditions. In an agile environment, however, major difficulties are encountered due to the cost and time required for each new part design to develop cause-and-effect relationships and to bring the process into specification. Without knowledge of the underlying error sources and their task-specific effects, it is difficult to apply the experience gained with one part to a different part.

The method developed in this project addresses the challenge of agile environments by monitoring trends in the machine tool errors instead of trends in the errors of similar parts. A relatively simple model is constructed that describes the accuracy of a machine. This error model does not necessarily rely on a detailed performance evaluation of the machine. Instead, operator expertise can be used to assign initial probability distributions to the various unknown parameters in the model. The resulting model captures our existing knowledge about the machine, while quantifying our lack of knowledge. Throughout the lifetime of the machine, any obtained information on its performance is used to tune the estimates for the various model parameters continuously through Bayesian regression. Examples of the respective data are inspection results of machined parts and results of quick periodic machine performance tests. Thus the machine

Exploratory Projects

Special Activities

error model learns from its mistakes and improves over time in areas important to the user. If the uncertainty of a prediction is too high, the model provides the justification for targeted machine performance measurements.

The project is a collaboration between the MEL's Manufacturing Metrology Division (MMD) and FTD. To this date, the following accomplishments include:

- Selecting a three-axis milling machine to study and validate the proposed method. The selected machine is used by FTD to machine a large variety of parts for its customers. We equipped the machine with temperature sensors at a few key locations.
- Defining a family of key error sources that will be studied in the project and identifying part features that are particularly sensitive to these error sources.
- Defining a monitoring protocol and start of data collection. The collected data on the machined parts includes their nominal geometry and tolerances, the process plan, tooling identification, operating conditions, time stamps, and inspection results obtained by FTD's quality department.
- Identifying an initial error model of the machine with assigned error distributions for key error parameters.

Uncertainties in machine error sources and in the operating conditions affect the machine performance

- Developing software modules to facilitate simulation studies; and
- Developing Bayesian regression procedures to tune estimates for the machine error parameters based on the inspection results of machined parts.
- Data for this machine will be gathered over the next several months. Final results and analysis of this method will be reported shortly after the conclusion of the project in May 2005.

Special Activities

Contact:
Charles McLean
charles.mclean@nist.gov

Concept Demonstration for Emergency Response Simulation

The environment at an emergency response site can get quite chaotic. People working the scene assume that someone is always in command and response is coordinated, however, that is not always true. Emergency responders are always training to keep up their skills and to learn how to different situations. Simulations can help meet this need. Currently, simulations are seen as a potentially useful training tool although the emergency responders to do not want to see it totally replace the hands-on exercises. As staff gets more exposure to advanced virtual reality technology, they may change their belief in how useful simulations can be in the future.

The objective of this project is to develop a concept demonstration prototype for the System for Integrated Modeling and Simulation of Emergency Response (SimsER), a system envisioned for bringing such capabilities to improve emergency response as conceptualized in the figure below. A demonstration prototype is sorely needed to help the emergency responders and potential stakeholders understand the value of the proposal. It will also help identify the difficulties involved in integrating such widely different simulations as a chemical plume model (continuous, computation intensive) and a traffic evacuation model (discrete, mid-level computation need).

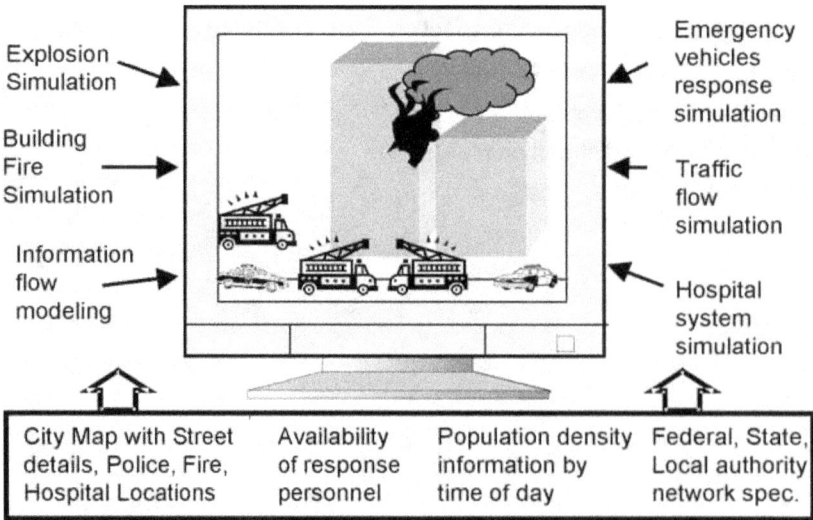

Special Activities

Current developers of modeling and simulation (M&S) tools mostly focus on developing their individual tool with proprietary inputs and outputs. They typically would offer to custom integrate their solution in user setups at a substantial cost. Such a mechanism is not suitable for most of the public jurisdictions or for use by this project. We propose an open approach based on standard data models to enable integration.

In the first phase of this project, a framework for modeling and simulation (M&S) of emergency response has been developed to define the scope and needed interfaces between M&S tools. Since a Geographic Information System (GIS) based tool is a necessary component for visualization of integrated outputs of simulation tools, interfaces and formats for GIS data have been studied. Capabilities and interfaces of data representation systems such as SEDRIS have been reviewed.

The team has also interacted with developers of M&S tools and potential user community to capture the requirements for integrated M&S of emergency response. The interactions include observation of two homeland security exercises involving use of simulation tools (Determined Promise'04 in Suffolk, VA, and Measured Response'04 at Purdue University), participation on a review panel for a project for development of such tools at U.S. Joint Forces Command, and meeting with representatives of San Francisco Emergency Operations Center.

Identification and evaluation of alternate architecture for distributed simulation for emergency response is also in progress. An initial evaluation of the current three candidates –High Level Architecture (HLA, developed by Department of Defense), Dynamic Information Architecture System (DIAS, developed at Argonne National Laboratory), and Synthetic Environments for Analysis and Simulation (SEAS, developed at Purdue University) has been completed. The applicability of game engine architecture as a front-end interface for distributed multi-participant simulation-based training is being evaluated. An emergency response scenario is being developed that will be used as the basis for concept demo and selection of the tools to be integrated.

Special Activities

Contact:
Ulf Griesmann
ulf.griesmann@nist.gov

A Geometry Measuring Machine (GEMM) for Metrology of Free-Form Optics

Precision optics require surfaces that deviate from their design form no more than a small fraction of the wavelength of visible light. Precision optics are the heart of the lithography process which is the foundation of the semiconductor industry. Industries as diverse as health care, aviation, defense, and electronics depend on ever more sophisticated precision optics. And yet, the efficient manufacture of precision surfaces, especially of those with non-spherical, or free-form, shapes, remains a serious challenge.

This project team plans to use the Geometry Measuring Machine (GEMM) as a universal metrology method for precisely measuring a new class of high performance optics with free-form surfaces. Free-form optics promise marked optical performance advantages but require a new approach to metrology. GEMM builds on the strengths traditional methods for metrology of free-form surfaces, interferometry and the coordinate measuring machine (CMM) offer, but avoids their liabilities. Like a CMM, GEMM measures local surface shape at every point of a survey grid on the part surface. Like full-aperture interferometry, GEMM uses an interferometer which it moves over the test surface. By minimizing the area of surface measured, GEMM minimizes the retrace error inherent with full aperture interferometry. For each GEMM measurement, parameters characterizing the local surface geometry are determined. For a one-dimensional profile this is simply the curvature. After calibration of the sensor, GEMM will measure surfaces of any geometry without test part specific null-optics. The measurement accuracy will match or exceed that of a CMM.

Special Activities

The key challenge in this project is the development of a 2-dimensional reconstruction method for the measurement of the whole surface of free-form optics. Another challenge is that the distance between sensor and part surface must be tracked to be able to correct the measurements for the residual retrace error of the micro-interferometer sensor. This capability will not be required to demonstrate the viability of the method but it may be necessary to achieve high measurement accuracy. The team developed novel ideas for solutions to both challenges, but a strong research effort is still required to validate them.

Solid model of NIST GEMMs Diagram

Although the primary deliverable of the project is a GEMM method for 2-dimensional surface measurement, a prototype instrument will be constructed. It will enable us to test the GEMM methodology with experimental measurements, and we will be able to compare GEMM measurements of actual parts with measurements made with interferometers. Numerical simulations of the GEMM method and the construction of the prototype instrument are currently under way.

Special Activities

Contact:
Nolan Brandenburg
brandenb@nist.gov

Microwire Electrical Discharge Machining for Fabricating Next Generation Precision Instruments

The objective of this project is to explore and evaluate the benefits and feasibility of utilizing micro-level wire electrical discharge machining technology for use in creating precision components, complex geometric mechanical parts and assemblies for various nanomanufacturing research programs in MEL and throughout NIST. This project team is exploring the level of precision, reliability and cost benefits that can be provided by this manufacturing technology and how it can be used to support the manufacturing service demands of MEL and other NIST research programs.

This exploratory project will assist the Fabrication Technology Division (FTD) in establishing expertise at NIST in Microwire Electrical Discharge Machining and in the engineering of complex geometries for devices used in nano-manufacturing research. With this expertise, FTD expects to increase the likelihood of expanding their manufacturing services to include this high-precision manufacturing technology for use in creating complex devices, instrumentation and mechanical assemblies needed throughout NIST.

A number of new NIST research programs have been established to address the anticipated needs of the U.S. nanotechnology and nanomanufacturing industry. To achieve the objectives defined in these research programs requires a technical approach that includes the creation of various complex instruments and/or micro-manipulation devices and assemblies. These new instruments and devices include complex geometric shapes and tolerance constraints that exceed the existing capabilities of machine tools and manufacturing processes provided by FTD.

Special Activities

To date, project research has included interfacing with microwire EDM manufacturers to assess the capacities, precision and the cost of available microwire EDM equipment. A study of micro and nano-machining processes in various industries is underway. This study will further define the market trends in private industry. Due to workflow in FTD, most of the research on this project will be completed in the Spring and Summer of 2005.

As NIST continues to create and measure smaller standards for industries, FTD's staff needs to be prepared to fabricate various testing apparatus and instrumentation on smaller and more precise scales. As a result of this project, capacities will be evaluated and requirements to support FTD's efforts in nanomanufacturing will be developed. As secondary benefit, the study allows FTD staff to evaluate available manufacturing technology for future capital improvement projects. These projects would ensure that the resources of FTD would match the needs of NIST.

Contact:
Maris Juberts
Maris.juberts@nist.gov

Analysis of Reference/Standard Test Facility Needs for Evaluating Performance of Intelligent Vehicle Systems

Intelligent vehicles are rapidly becoming an important component of the US military and of the transportation industry. Unmanned Ground Vehicles (UGVs) are used in mine clearing and search and rescue operations around the world and major defense programs are gearing up to place sophisticated autonomous UGVs in the soldier's hands by 2015. The Department of Defense (DOD) plans to use experimentation environments and UGV performance measurements to specify functionality, to measure progress/maturity and to direct research. In the transportation sector, US automotive manufacturers and the US Department of Transportation (DOT) are striving to improve the safety of cars by using advanced sensors and intelligent driver assist systems to reduce collisions on roads. Increasingly, US manufacturers are using intelligent crash warning systems as a major factor in competing with foreign car manufacturers. The US National Highway Traffic Safety Administration (NHTSA) is charged with setting safety performance standards and relies heavily on test specifications, test data and rating criteria. Performance measurement will help facilitate the insertion of improved safety features on automobiles.

The main objective of the exploratory project is to strengthen MEL's and NIST's ability to meet the nation's metrological needs pertaining to performance measurement of intelligent vehicles. This effort includes investigating potential Other Agency program areas where NIST may contribute their experience and knowledge to developing standardized measurement methods and procedures towards the objective of evaluating the performance of Intelligent Vehicles systems and subsystems performing realistic operational scenarios.

Several DOD programs are developing unmanned ground vehicles that implement intelligent behaviors. Critical to the success of these programs is an experimental environment to speed development, lower costs and measure progress. One of the DOD programs – The Joint Robotics Program (JRP) is starting a program called the National Unmanned Systems Experimentation Environments (NUSE2). Staff members from this project proposed a formal collaboration (submitted September 30, 2004) between NIST/MEL and the NUSE2 group to develop a system that would specify, characterize, model and certify outdoor experimentation environments that could be used for the evaluation of intelligent autonomous and semi-autonomous robots and intelligent

Special Activities

behaviors.

MEL's staff will work with NUSE2 members and other interested parties to develop a comprehensive assessment program for fostering the development and evaluation of autonomous navigation performance of small robots. Current robot technology requires tedious control by the operator for low-level tasks such as driving to goal points specified by a user. Based on our experience and knowledge, we believe vendors will be offering robots with path planning and obstacle avoidance capabilities within a few years. Therefore, the project team's initial effort will directed to measure the performance of autonomous and semi-autonomous navigation, which includes obstacle detection, hazard avoidance, route planning and execution. This effort will lay the foundation for assessing performance of other navigational modes of operation, manipulation tasks and tactical missions.

Working with NUSE2 and other domain specific experts, our group will select, and analyze and translate representative missions into a sequence of tasks based on a hierarchical task decomposition methodology. The tasks will then be used to develop specifications for an Intelligent Behavior Assessment Course (IBAC). Using the performance metric parameters generated by a vehicle's path and response to the course features, the group will be able to assess a vehicle's performance.

In addition to the IBAC design, the project team will develop methods for measuring ground truth, site characterization and for capturing robot performance data. Ground truth will quantify system performance and populate simulation environments. Robot performance data is primarily position

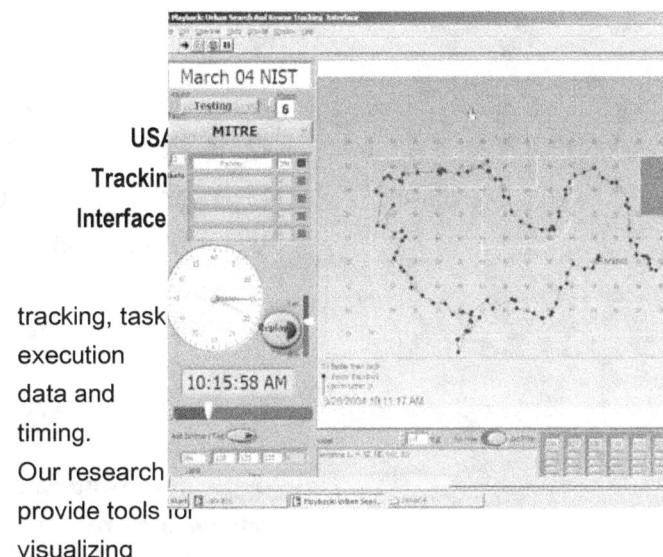

tracking, task execution data and timing. Our research provide tools for visualizing task execution, maintaining quality control and publishing results.

In addition to the work with the DOD, our group met with Dr. August Burgett, technical director of the NHTSA as part of the Intelligent Vehicle Initiative for the DOT. Dr. Burgett expressed a need for determining "ground truth" high accuracy, real-time measurement of relative and absolute position and velocity between an instrumented test vehicle and other moving vehicles and objects on a road. The need is for a naturalistic (in traffic driving) approach for evaluating the performance of vehicle-based safety systems that use advanced sensors to reduce vehicle accidents/collisions on roads.

NHTSA and the Intelligent Transportation Systems program office approved funding for our participation in new program planning for future initiatives in their Intelligent Vehicle Initiative program. Funding for NIST in FY2005 is expected to be at least $350 K and is expected to continue for four to five years until Field Operational Tests are completed. We will develop a next generation measurement system to collect data for independent evaluation of the performance of Integrated Vehicle-based Safety Systems.

Contact:
Kari Harper
Kari.harper@nist.gov

Quadrature Laser Interferometry for Dynamic Displacement Measurements in Vibration Metrology

Quadrature laser interferometry (QLI) can be used to perform dynamic displacement measurements by primary methods and has many advantages over the conventional Michelson interferometry currently used in our accelerometer calibration services. **Implementation of a quadrature laser interferometry system allows absolute (primary) calibrations of accelerometers over an extended frequency range, at a continuum of displacement amplitudes within that range, and provides means for measurements of complex sensitivity (magnitude and phase) of accelerometers.** Efforts have been made by a number of National Metrology Institutes (NMIs) to implement QLI in vibration metrology with varying degrees of success. Implementation of QLI presents many challenges, e.g., hardware configurations, sampling rates, and signal processing algorithms, which all influence the overall uncertainty of the measurement system.

It is no secret that the advance of technology demands that MEL continue to expand its measurement capabilities. This is certainly true in the area of vibration metrology. Consider, for example, machining technology. As spindle speeds increase to levels unimagined a few years ago, vibration frequencies of interest for machine tool and process performance assessment and modeling increase as well. Another trend increasing the importance of higher frequencies is the miniaturization of systems due to the higher resonant frequencies of such systems.

The initial implementation of a quadrature laser interferometry-based system for dynamic displacement measurements is being done on the NIST Super Shaker. Then, based on competencies gained in this exploratory project, similar improvements can be obtained in other measurement services equipment currently in use, extending the range of calibration frequencies we can offer our customers and providing them with phase data in addition to the magnitude of the accelerometer response to vibration.

Special Activities

To develop a top performing system based on quadrature interferometry, a study of the state of the art of implementations by other NMIs has been done. On the basis of this study, the project team decided to use homodyne, rather than heterodyne QLI for the first implementation. This decision reflects both the potential benefits of a homodyne system, its relative ease of implementation as compared to a heterodyne system, and our budgetary constraints.

A review has been done of recent advances in the optical, electronic, and signal processing hardware and algorithms that can serve as components of a QLI system. We carried out a comparative performance analysis of available hardware and signal processing algorithms required for such a system. An initial set of hardware components has been selected that meet both our budgetary and technical requirements. These have been procured, and elements of the prototype system are currently either under final design, fabrication or assembly.

It is expected that the remainder of the project will be devoted to completion of the prototype QLI system and integration of that system on the NIST Super Shaker.

Special Activities

Recently Completed Exploratory Projects

Contact:
Jack Stone
Jack.stone@nist.gov

Precise Length via Satellite: GPS-based Optical Frequency Combs for Length Measurement

The goal of this project was to determine the feasibility of using timing signals received from the Global Positioning Satellite (GPS) system, along with optical frequency-comb technology, to provide new wavelength standards for measurement with very direct traceability to the definition of the meter. In the course of this project the project team developed and demonstrated the use of a prototype GPS-based system for calibration of laser vacuum wavelength, identified the important areas where opportunities exist to simplify or improve the system and thus make it more practical, and began looking at possible new applications of the system. Preliminary results have been encouraging and suggest that this technology will have significant impact on length metrology at the National Metrology Institute (NMI) level. More widespread impact in industry is also possible, but cost and complexity issues remain a barrier; addressing these issues is a continuing need. The work begun in this project will be continued through a NIST Competence Project starting in FY2005.

Optical frequency combs are a breakthrough technology that can provide precisely know frequencies (or vacuum wavelengths) throughout the visible and infrared spectrum. The comb is a band of optical radiations at equally spaced frequency intervals, generated by a mode-locked femtosecond laser. The frequency spacing of the comb components is set by the pulse repetition frequency of the femtosecond laser. Knowledge of the repetition rate, along with knowledge of an offset frequency, determines the frequencies of comb components. Thus optical frequencies at hundreds of terahertz can be accurately related to precise measurements of the repetition rate (typically hundreds of megahertz).

Special Activities

For the last twenty years iodine stabilized lasers have served as the primary transfer standard in linking the definition of the second to length metrology. For example, the formal definition of the meter is the length of the path traveled by light in vacuum during a time interval of 1/299 792 458 of a second. Today combs are beginning to replace the iodine stabilized laser as the pivotal link between frequency and length. The optical comb technology provides greater accuracy, with a very direct tie to the definition of the meter, and is infinitely more flexible than previous approaches. Almost any visible or near-infrared laser wavelength can be calibrated using a comb, whereas a particular iodine stabilized laser provides only one of a few possible standard wavelengths. The comb is still much more complicated than an iodine stabilized laser, but if more than one wavelength reference is required (for example, for multicolor interferometry) then the comb is a logical alternative to the iodine stabilized laser.

An optical frequency comb plus a "clock" thus provides an attractive new mechanism for realization of wavelength standards. Furthermore, the clock need not be a local cesium standard; more attractive for our purposes are to make use of timing signals from the Global Positioning System. The GPS signal, averaged for 1 day, has a frequency uncertainty below 10^{-12}, more than an order of magnitude better than the uncertainty of the iodine stabilized laser. Thus, GPS combined with the frequency comb provides a mechanism for traceable, world-wide dissemination of frequency/wavelength standards with verifiable uncertainty.

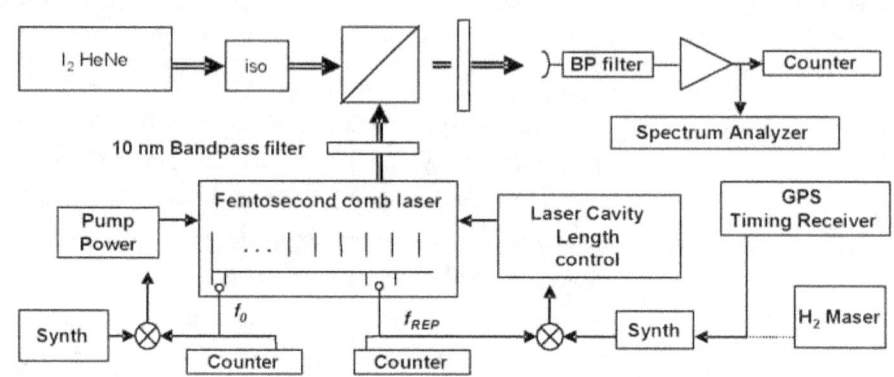

Setup for calibrating Iodine-stabilized HeNe laser with comb.

Special Activities

Contact:
David Stieren
david.stieren@nist.gov

Fuel Cell Manufacturability

This exploratory project team identified the challenges associated with fundamental manufacturing technology for polymer electrolyte membrane (PEM) and solid oxide fuel cells. The team sought to identify areas of MEL core competency that could be leveraged to address fuel cell manufacturability issues, highlighting measurement, standards, and infrastructural technology areas. Within scope were manufacturing-related issues and technical challenges associated with fuel cells for transportation (especially automotive), stationary power, and portable power applications.

The project team researched the fundamentals of fuel cell technology and engaged stakeholders from the community to assess the state of fuel cell manufacturability. This included sponsoring a fuel cell manufacturability workshop in December 2003 in Dearborn, Michigan, that brought together a broad national audience to: (1) identify critical manufacturing issues associated with high volume fuel cell production; (2) explore the development of a broad, national strategy for fuel cell manufacturability; and (3) provide input to the identification of an MEL role in the area. Representatives from over two dozen fuel cell organizations from industry, government, and academia attended the workshop, confirming that manufacturing plays important role in making fuel cells more affordable and is on the critical path to commercialization.

The fuel cell industry is still very much emerging. This project confirmed that many technical challenges exist relating to the manufacturability of fuel cells, including the need to address manufacturing issues while fuel cell designs are evolving. Low-volume, high-cost prototype and laboratory fabrication methods must be transformed into full-scale, high-volume, low-cost manufacturing processes. A critical enabler here is the development of the requisite metrology to facilitate the control of manufacturing processes and products.

Special Activities

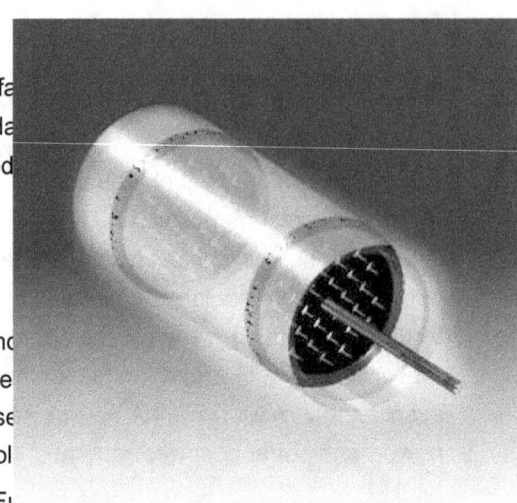

The project uncovered sufficient manufacturing issues and measurement / standards needs for fuel cells to warrant continued MEL involvement in the area. Project recommendations for MEL included:

• continuing to represent NIST on the Hydrogen and Fuel Cell Research and Development Interagency Task Force being coordinated by the White House Office of Science and Technology Policy;

• continuing to participate in the NIST Fuel Cell Working Group; and

• continuing to be involved with the fuel cell community through events, seminars, and colloquia to identify collaboration and resource acquisition opportunities for MEL.

Special Activities

Contact:
Richard Silver
richard.silver@nist.gov

New Advanced Optical Methods for CD and Overlay Metrology Beyond the 65 nm Node

Although optics are often thought to lose its effectiveness as a metrology tool beyond the Raleigh criterion, there is now evidence that optics can be used to image and measure features as small as 10 nm in dimension. Fundamental imaging limits presented a real concern for optical-based measurements. Optical methods are used extensively for level-to-level overlay metrology, defect inspection, and applications that require critical dimension (CD) metrology. With numerous semiconductor manufacturing applications and increased growth in nanomanufacturing, the possibility of using high-throughput, extremely sensitive, cost effective optical metrology methods is very appealing.

As demonstrated by results obtained in this exploratory research project, optical imaging and metrology with resolution well beyond the wavelength is now a possibility using a new optical configuration and design, increased computation speed, high-resolution Charged-Coupled Device (CCD) array technology, improved signal processing of large arrays, and reliable theoretical scattering models. The essential optical components involve low numerical aperture optics, multi-pole illumination schemes, and access to the pupil plane resulting in structured high angle to very low angle illumination. Another key technological advance is the realization that the through-focus phase information, accessible by imaging through focus, provides a sensitive, complex optical signature for a given sample and optical configuration. The sample or target design also plays a central role in the ability to increase optical sensitivity. A target structure, which enhances the amplitude of the scattered fields and the desired information content, in concert with the optical configuration, can be optimized for this purpose. The importance of a structured target has also been shown in this study to play a substantial role in final image formation and increased sensitivity at the nanometer scale.

Scattering calculations and experiments imaging features as small as 40 nm wide silicon line indicate this technique has very real promise for success. The calculations show the signal to noise and detection of features smaller than 10 nm to be possible and even potentially practical. Actual optical images were acquired with good signal to noise ratios for silicon features 40 nm in lateral dimension. To investigate the potential of these new optical methods we have performed an extensive set of simulations and experiments. Results summarizing the technical

Special Activities

gains with this methodology are shown below as well as data which explores the fundamental limits of model based optical imaging using the full scattered fields.

What aspect of the investigation focused on low Illumination Numerical Aperture (INA) imaging methods, similar to scatterometry, except using a more conventional optical microscope layout. This has significant potential to enable full field imaging of scattered optical fields with sensitivities to high diffraction orders. The potential gains from using low numerical aperture illumination in combination with model based analysis and interpretation are shown below.

The electromagnetic scattering models used in this study have been developed in-house and in collaboration with Spectel Co. Optical modeling, which NIST is considered an international leader, plays a fundamental role in optical imaging at this scale

A graphic representation of nanotube based devices as might be encountered in future manufacturing

Potential Impact

This work completed in this exploratory project has the real potential to impact high-resolution optical metrology. This methodology has not been previously explored, and relies on advanced high-speed computation and new methods for optical illumination and low NA imaging. The results achieved here are already having an impact as other research and development teams are now implementing these techniques.

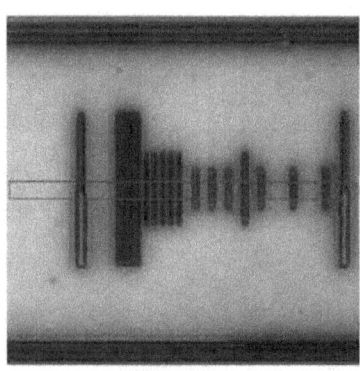
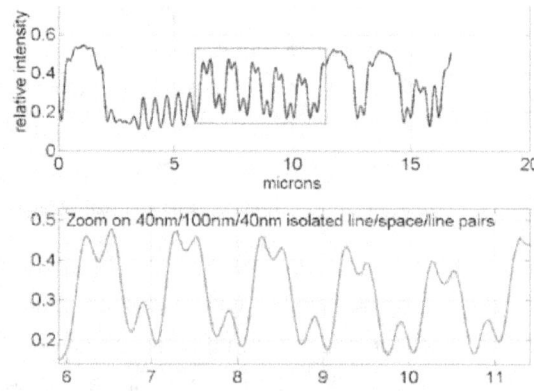

An experimental demonstration of the imaging capabilities at 100x magnification. 40 nm CD targets are shown on the left with imaging wavelength of 436 nm. The upper right panel in highlights profiles of a line/space/line array consisting of two 40 nm lines separated by a 100 nm space at a pitch of 1 micrometer

Special Activities

Contact:
Ram Sriram
ram.sriram@nist.gov

Healthcare Information Interchange Through Shared Ontologies

In today's medical practice, patient health information is codified using specialized healthcare terminologies. The uniformity of representation through codes (as opposed to free form natural language statements) supports a variety of data processing functions as well as information between medical record, laboratory and diagnostic imaging systems. The healthcare terminologies in use today include Systematized Nomenclature of Medicine (SNOMED), Unified Medical Language System (UMLS), and Read Codes. Data interchange between healthcare institutions is supported through a number of standards including Digital Imaging and Communication (DICOM) for diagnostic images, Health Level 7 (HL7) for components of electronic medical records, and American National Standards Institute (ANSI) X12 for healthcare business information interchange. Transactions that contain clinical information, such as patient referrals between healthcare institutions, carry coded medical information that can be mapped back to the organizations' local medical information representation. At present there are two major barriers to the full range of potential medical data interchange: 1) there are large disparities among the terms used and the codes generated by the various healthcare vocabularies; and 2) there are no semantics associated with the terms in the vocabularies that would allow the recognition and reconciliation of different terms that have the same meaning. These major barriers present a technical challenge that intrigued our project team. They proposed that the use of shared ontologies (i.e., a formal description of the concepts and relationships that can exist between them) would resolve semantic ambiguities and harmonize dissimilar information models of patient care information in collaborating healthcare practices.

In the past year, the project team:

1. Established contacts with key healthcare informatics standards developme organizations (SDOs) and healthcare organizations.

The group pursued collaborations with: the Cleveland Clinic, the Union Hospital located in Cecil County, MD, the Health Level 7 standard organization and the National Institutes of Health. The group participated in the planning and organization of several workshops related to data interoperability in clinical settings.

2. Identified requirements for healthcare information interchange for specific scenarios.

Referrals to specialists and referrals for diagnostic imaging were the two scenarios, which involve complex healthcare information flows, were identified as those having potential for computer-assisted resolution of ambiguities in terminologies and semantics. The group gathered requirements for healthcare information interchange from the technical literature that provides workflow and information flows for radiology and referrals. Unfortunately, the practices at the collaborating healthcare organizations reduced the scope of work: Cleveland Clinic primarily conducts referrals within their system; while Union Hospital focuses on inpatient care and does not use referrals; moreover both organizations employ proprietary homogeneous healthcare information systems.

3. Identified ontologies relevant to the prototype scenario.

For the radiology referral prototype scenario, the ontologies will accommodate the DICOM Structured Reporting (DICOM SR), the Radiological Society of North America's (RSNA) RadLex terminology, as well as HL7. They considered information models such as HL7's Reference Information Model (RIM) and process models underlying clinical guidelines such as Guideline Interchange Format (GLIF), Stanford's EON Guideline Model and the Object-oriented Guideline Expression Language (GELLO).

4. Proposed an architecture for the application of semantic mapping to healthcare data interchange standards.

The proposed architecture of the Shared Ontology Service is a hybrid encompassing the Mapping Ontology Service approach of Stanford University and the common ontology approach (e.g., UMLS); integrated with an HL7-compliant Terminology Service and an open interface to facilitate integration with other Ontology Services or terminology services. The architecture will enable mappings at various levels: activities/subactivities; inputs/outputs; resources; and flow constraints - determined by the underlying metamodel.

5. Identified ontologies and semantic mapping techniques for the selected scenario.

The group used the UMLS Metathesaurus to create semantic mappings. Concepts and associated expressions in the UMLS Metathesaurus are used to create semantically equivalent translations while Parent/Child and Broader Term and Narrower Term (BT/NT) relationships are used to create non-equivalent relationships between terms.

The work from this project will continue under the Manufacturing Metrology and Standards for the Health Care Enterprise program that was started in FY2005.

Special Activities

Contact:
Victor Nedzelnitsky
vnedzelnitsky@nist.gov

Measurements For Improved Communications And Listening Equipment Effective In Noisy Environments

Identifying the source and type of sounds can be quite critical to the U.S. military. For example, identifying the click of a bolt-action rifle of a sniper preparing to fire could mean the difference between life and death. The enhanced ability to detect, classify, recognize, and understand sound signals that must be received relies on more advanced techniques than conventional amplification alone. Security and military applications would like to be able to identify acoustic signatures and track ground vehicles, fixed and rotary-wing aircraft, and cruise missiles.

In the future, advances in electronics, signal processing, the design and fabrication of miniature microphones and, particularly, arrays may allow new microphone and earphone systems to provide improved real-time detection of signals such as speech and acoustic signatures in the presence of interfering noise and reverberation, and even raise the possibility of "super-hearing" capability.

New and evolving equipment claims to enhanced communication, hearing, and acoustical signal detection and tracking. These features are needed for emergency response, security, and military personnel in acoustically noisy environments. However, critical measurement capabilities still need to be developed. These measurement capabilities are key prerequisites to the development of relevant protocols, performance data, and standards essential to the evaluation, improvement and optimization of equipment for Homeland Security and U.S. military applications. These applications range from communication, intrusion detection, identification, tracking, and targeting of hostile personnel and vehicles, to locating the injured. This project team began laying the foundation for a measurement service by investigating the necessary critical enabling measurement capabilities and researching applicable design, manufacture, and tests of new acoustical array components and systems.

Special Activities

Using NIST's large anechoic chamber, the team established the importance of measurement capabilities to determine free-field directional patterns, beam patterns, and inter-microphone transfer functions of arrays designed with desirable characteristics for detecting signals in the presence of interfering sounds. They modified and improved their current equipment to develop an interim four-channel capability for measuring inter-microphone transfer functions and higher-resolution polar response patterns. The team collaborated with Planning Systems, Inc. (PSI) to: a) complete the NIST acoustical measurements that support the U.S. Army's Acoustic Counter Battery System Technology Demonstration program. This program is an initiative to demonstrate a non-line-of-sight covert acoustic targeting capability to better exploit the acoustic environment of the battlefield; and b) perform customized measurements including free-field directional patterns, beam patterns, and inter-microphone transfer functions for prototypes of a classified directional array sponsored by Department of Defense (DOD). One of the project team members, Vic Nedzelnitsky was invited and participated in a Defense Sciences Research Council (DSRC) and Defense Advanced Research Projects Agency (DARPA) study meeting on new technologies (including acoustical technologies) and possibilities for human sensory modulation for jungle warfare.

The project team plans to take advantage of opportunities in the future to attend meetings similar to the DSRC/DARPA meeting to develop promising, mutually beneficial interactions with colleagues, customers, and partners. They plan to continue working with PSI and other customers to further develop our capabilities in this evolving area of acoustic measurement science. The availability of improved communications and rescue equipment will benefit first responders (i.e., firefighters, rescue and emergency medical service personnel, and police), command and control personnel, National Guard personnel, and the public they protect. Improved communications capabilities, efficiency, and coordination among law enforcement personnel and emergency responders will save lives and money, will better protect secure areas, and will support forensic investigation and prosecution.

Special Activities

The National Science and Technology Council
Interagency Working Group on Manufacturing Research and Development

Contact:
David Stieren,
david.stieren@nist.gov

Background

MEL plays a lead role in the newly formed National Science and Technology Council (NSTC) Interagency Working Group (IWG) on Manufacturing Research and Development. The IWG was chartered in 2004 in response to the Commerce Department report, "Manufacturing in America," to identify and integrate requirements, conduct joint program planning, and develop joint strategies for the manufacturing research and development (R&D) programs conducted by the Federal government. MEL is the NIST representative to this body, which is chaired by the Commerce Department Under Secretary for Technology. The IWG vice chair is the director of MEL; the MEL strategic relations manager also has a key role in this effort.

The NSTC was established by Executive Order in 1993 to coordinate the diverse parts of the Federal R&D enterprise. Chaired by the President, the NSTC includes the Vice President, the Assistant to the President for Science and Technology, and Cabinet Secretaries and Agency Heads with significant science and technology responsibilities.

The NSTC is charged with setting national goals for Federal science and technology investments in areas ranging from information technologies and health research to improving transportation systems and strengthening fundamental research. To support these goals, the NSTC prepares R&D strategies that are coordinated across Federal agencies. Within the NSTC, the IWG is a forum for developing addressing issues associated with manufacturing R&D policy, programs, and budget guidance and direction. Designed to promote information exchange among its participating agencies, the IWG builds upon the activities of the Government Agencies Technology Exchange in Manufacturing (GATE-M).

Special Activities

Interagency Working Group on Manufacturing Research and Development

Established and led by MEL, GATE-M put in place mechanisms to: (1) enable the member federal agencies to exchange information about their manufacturing technical programs; (2) coordinate manufacturing R&D programs among federal agencies to facilitate collaboration; and (3) provide a forum for the agencies to advocate for manufacturing issues on an interagency, national level. The MEL Director chaired the GATE-M effort and the MEL Strategic Relations Manager served as the GATE-M executive secretary.

The following agencies participate on the IWG:

- Department of Agriculture
- Department of Commerce / NIST
- Department of Defense
- Department of Education
- Department of Energy
- Department of Health and Human Services / National Institutes of Health (NIH)
- Department of Homeland Security
- Department of Labor
- National Aeronautics and Space Administration (NASA)
- National Science Foundation (NSF)
- White House Office of Management and Budget (OMB)
- White House Office of Science and Technology Policy (OSTP)

The functions of the IWG are to:

- propose policy recommendations for manufacturing R&D
- engage in interagency manufacturing R&D program planning and budgeting
- review agency priorities and technical issues by providing a forum for agencies to exchange program-level information about their manufacturing R&D activities
- promote communications among the government, private sector, and academia on R&D requirements and programs
- identify opportunities for collaboration, coordination, and leverage among agencies in specific technical areas
- report and make recommendations to the Committee on Technology and to OSTP regarding Federal manufacturing R&D priorities and the need for specific interagency activities to address those priorities

While the IWG is focused on coordinating Federal manufacturing R&D efforts, it will work with the private sector to ensure that Federal activities are well aligned with the needs of industry. The IWG may seek advice from members of the President's Council of Advisors on Science and Technology (PCAST) and from the private sector broadly, in a manner consistent with the Federal Advisory Committee Act (FACA).

Special Activities

Initial Results

In 2004 the IWG identified three manufacturing R&D priority areas. Using a priority list developed earlier by GATE-M and adding other priority topics at their discretion, each agency identified its top three or four priority topics. The IWG narrowed down this list on the basis of national need, the need for interagency collaboration, and the potential for becoming a Federal budget priority. The IWG technical priority topics are being developed as the basis of an IWG recommendation for a future coordinated, multi-agency budget initiative in manufacturing R&D.

The three priority areas are:

- Intelligent and Integrated Manufacturing Systems
- Manufacturing for the Hydrogen Economy
- Nanomanufacturing

Interagency task teams chartered by the IWG are preparing white papers that define the challenges and technical approaches to them in each of the three areas. MEL leads the Intelligent and Integrated Manufacturing Systems team and participates on the other two; all three priority areas are well aligned with MEL's programs.

The high-level goal of the manufacturing R&D initiative is to lead the development and implementation of advanced manufacturing technologies in these key technology areas for the U.S. manufacturing sector to benefit the U.S. economy. A secondary goal is to increase the effectiveness of the overall Federal manufacturing effort by improving planning, coordination, and collaboration among Federal agencies in these areas.

The IWG findings and recommendations will be discussed with interested parties in the private sector, including industrial associations and others with a strong interest in manufacturing research and development, to obtain their advice on and their support for the proposed work. After any resulting changes, the IWG white papers will be consolidated into a report that will be submitted to the NSTC in early 2005 for publication.

Contact:
Clarence Johnson

ceejay@nist.gov

FTEs:
1.0
(program administration)

Annual Program Funds:
$370K

Intelligent Manufacturing Systems (IMS)

Program Goal

The goal of the Intelligent Manufacturing Systems (IMS) program is to develop, in a coordinated worldwide effort, the next generation of manufacturing and processing technologies. The primary work of the program is carried out through a series of industry-led international research and development (R&D) projects. The U.S. Secretariat that facilitates the participation of U.S. entities in the IMS program is located in MEL.

Introduction

IMS grew out of an initiative from Japan proposed in 1989 by Professor Hiroyuki Yoshikawa, then President of the University of Tokyo. His vision for IMS was for a global system of industrial cooperation and technology sharing for the benefit of mankind and in particular, for the benefit of entities involved in IMS projects. The formal ten-year IMS program began in 1995 following a two-year feasibility study (1992-1994) and is scheduled to conclude in April 2005.

Technical Approach and Management

Each IMS project must have participants from no less than three regions. Entities from all seven IMS regions – Australia, Canada, the European Union and Norway, Japan, Korea, Switzerland, and the United States – are eligible to collaborate on IMS projects and activities.

IMS has a support structure for conducting projects that provides specific arrangements for the protection of intellectual property rights (IPR) and enables small enterprises to cooperative effectively and on equal footing with larger ones. The ability to work cooperatively with leaders of technology and industry to share costs, risks, and expertise is a necessity in today's global environment. IMS is the only major international research program that allows participants to tap into a well-established global network of manufacturing innovation to help

Special Activities

improve manufacturing operations, enhance international competitiveness, and lead to technology breakthroughs via market-driven research and development (R&D). It provides access to technology that might not be available within a region. The market-driven character of IMS R&D means that academics and researchers are part of an effort to transfer valuable technologies to future generations.

Five technical themes provide the broad framework within which IMS operates. All projects must address one of more of the following technical themes:

- Total Product Life Cycle,
- Process,
- Strategy/Planning/Design Tools,
- Human/Organization/Social; and
- Virtual/Extended Enterprise.

IMS provides guidelines for preparing and submitting project abstracts and proposals. All regions review each submission for adherence to those guidelines and the technical themes. The average length of an IMS project from beginning to completion is three years. Currently more than 550 companies and research institutions including over 100 from the U.S. are active in IMS research consortia.

The IMS International Steering Committee (ISC) comprised of delegations from the member regions manages the program according to the Terms of Reference, the rules by which IMS governs itself. Mr. Robert Cattoi, Rockwell International (retired), chairs the ISC and Mr. Robert Kiggans, President and Chief Executive Officer of ATI, is Head of the U.S. Delegation. The IMS Inter Regional Secretariat (IRS) currently located in Washington, DC, coordinates the technical activities of IMS with the assistance of a Regional Secretariat in each of the IMS regions such as the one for the U.S. located in MEL.

Special Activities

Impact of Phase I

During Phase I of the IMS program –

- 16 projects have been completed; the U.S. participated in seven of them,
- 16 additional projects are ongoing; some are the second phase of original projects,
- Three full proposals are awaiting endorsement pending completion and signing of their Consortium Cooperative Agreements,
- 14 abstracts have been endorsed and will become full proposals shortly,
- 19 outline proposals are in circulation and are looking for additional partners,
- Over 300 million research dollars have been expended to date, and
- Estimated expenditures for this year are expected to be $55 million.

To date, 715 partners have been involved in IMS-endorsed research. As the organization is industry-led, over 70 % are industry partners. Last year's survey of all IMS projects reported that 100 % of the project teams felt they had increased mutual trust with their project partners, 96 % gained valuable experience in international collaboration, 96 % reported that their projects were and continue to be successes, and 83 % reported that IMS provides an "excellent" framework for project success. With 42 patent applications filed and 13 granted, 181 copyrights generated, ten licenses issued for background and foreground, and presentations given at over 1300 meetings and symposia. IMS is already a proven program.

In addition to the projects generated during Phase I, IMS sponsored other activities such as the **Vision 2020 Forum** held in Irvine, California, in 2000 that identified a concept of "new manufacturing" operating as a service provider industry with new systems of corporate architecture and highly integrated enterprise functions. Forum attendees identified ten key issues for global manufacturing in 2020. The event was by invitation only and was attended by representatives from all IMS regions.

The IMS Project Forum 2001 held in Ascona, Switzerland, focused on the dissemination of the technical and developmental efforts of the participating partners of the IMS program. Over 100 participants attended this event and twenty-five active projects presented their activities. The Forum provided an excellent platform for the IMS program partners to present their research and development activities and exchange their experiences and ideas.

The **IMS Forum 2004** held in Lake Como, Italy, in March, 2004 attracted 460 attendees who analyzed trends and needs in manufacturing, disseminated the scientific results of the research projects, stimulated new projects and activities and provided IMS projects with an opportunity to have project meetings. The attendance and results from this Forum helped convince IMS management that a second phase of the program was warranted.

Plans for Phase II of the IMS Program

Based on the successes of Phase I of IMS and the desire of member regions to continue the program, the ISC is currently planning for Phase II. A working group has completed a new Terms of Reference (ToR) and this document is currently being submitted to the governing authorities of the IMS regions for approval. In the case of the United States, that document will be submitted to NIST (the National Institute of Standards and Technology) for legal review. Once three regions have approved the new ToR, Phase II of the program can begin. At the present time, all current member regions have indicated an interest in participating in Phase II and new member regions are expected to be added. Korea will serve as the first chairman of Phase II and will provide the facilities for the Inter Regional Secretariat.

The IMS ISC has identified the following four goals for Phase II of the program:

1. Promote and strategically lead the advancement of knowledge in manufacturing through various IMS products and services. The #1 priority for IMS is to continue to foster successful IMS projects.

2. Advance the esteem of manufacturing and in so doing, establish the IMS brand image.

3. Encourage new-region involvement.

4. Support global manufacturing education and training.

IMS is the first worldwide program to address manufacturing challenges and sustainable production systems in the 21th century. Its projects are vital in shaping manufacturing today and through its current projects and established research networks, will continue to influence manufacturing science in the future.

For further information including guidelines for project formation, check the IMS website (www.ims.org).

Contact:
Simon Frechette
simon.frechette@nist.gov

Systems Integration for Manufacturing Applications (SIMA)

Customer Need

The need to share computer-interpretable information among different software systems is very nearly as old as computing technology itself. Among today's information technology practitioners this need is typically referred to as that of interoperability between software systems.

When software systems lack an inherent capability to share information among one another we say they are not interoperable. When businesses require software systems to be interoperable and they are not, problems ensue. These problems are typically manifested as data transfer errors and the extra staff time required to manually fix or re-enter data from one system into another. The costs resulting from such interoperability problems are high; annual costs of $1 billion per year were documented in analysis of interoperability costs in the automotive sector. Private sector estimates were put this figure at $20 billion per year.

Lack of desired interoperability between software systems can be found in every industrial sector, in every software market, among different versions of the same software, among identical versions of the same software on different platforms, and even among identical versions of the same software on the same platform reflecting conflicting usage of optional settings. NIST has partnered with industry to address a variety of software interoperability problems. NIST's efforts in the solution of software interoperability issues are typically manifested as contributions to the technical content of voluntary standards through participation in standards developing organizations; development of reference software implementations of voluntary standards; and development of procedures for testing software that implements voluntary standards.

Special Activities

Such efforts have led to the adoption of software interoperability solutions like the Structured Query Language (SQL), the Initial Graphics Exchange Specification (IGES), the Standard for the Exchange of Product model data (STEP) and new business communication standards such as ebXML. These and similar NIST efforts are increasing the interoperability of software systems and resulting in concomitant higher labor productivity. Still, increasing numbers of software interoperability solutions are being sought for a multitude of industries. There are multiple factors contributing to this growth in the need for software interoperability solutions including outsourcing of business functions, acquisitions/joint ventures, increasing information technology (IT) deployment, and increasing IT capabilities.

Technical Approach

SIMA supports technical efforts throughout all of NIST's laboratory organizational units. These efforts contribute to the definition and creation of standards facilitating the exchange of manufacturing data within and across all levels of the product realization process. Objectives include the establishment of:

- Rigorous methods for defining and testing interoperability solutions.
- Standards specifying information to be exchanged as well as the interface mechanisms necessary to do so.
- Tests validating potential standards solutions and implementations.

There are four principal reasons why industries seek NIST's involvement in solution of interoperability problems:

- Neutrality – NIST has no vested interest in the interoperability solutions being promulgated.
- Rigor/integrity/expertise – NIST has extensive domain knowledge combined with an established track record of contributions to interoperability solutions.
- Breadth – NIST provides a multi-industry perspective.
- Permanence – NIST has a long-term perspective that is not subject to the whims of short-term market dynamics.

Special Activities

SIMA Program Office Activities

The SIMA program focuses on the development of domain-specific standards specifying information to be exchanged or shared or the interfaces among systems; rigorous methods to define/test interoperability solutions; performance of tests validating potential standards solutions and implementations; standard reference data, and internet-based mechanisms providing unique NIST resources to industry practitioners. The SIMA program provides unique NIST facilities and coordination of cross-domain activities. SIMA customers include Aerospace, Automotive, Bio, Chemical, Electronics, Engineered Materials, Industrial Equipment, Pharmaceutical, Software Vendors, Construction, Genomics, and Standards Development Organizations.

2005 Projects
- Numerical Data Markup Language - UnitsML
- Infrastructure for Integrated Electronics Design and Manufacturing
- Adaptable and Automated Testing Tools
- Visualizing Ontologies
- Interoperability of Databases for Inorganic Materials
- Data Standards for Structural Bioinformatics
- Development of Mathematical Content Semantics
- eBusiness Standards Convergence - eBSC
- Interoperability Standards for Capital Facilities Equipment
- Analytical Chemistry Data Interchange Standards
- Physical and Chemical Property Data Interchange Standards
- Integration of Building Environment Information
- Virtual Reality Standards & Integration of Anthropometric Data
- Infrastructure for Neutron Reflectivity Component Integration

NIST Cost Studies Available

Cost studies directed by the SIMA program have proved extremely valuable to MEL and NIST. These include:

- Interoperability Cost Analysis of the U.S. Automotive Supply Chain, 1999 (MEL)
- Economic Impacts of Inadequate Infrastructure for Software Testing, 2002 (ITL/MEL)
- Economic Impact of STEP in Transportation Equipment Industries, 2002 (MEL)
- Strategies for eManufacturing Standards (MEL), 2003
- Economic Impacts of Inadequate Infrastructure for Supply Chain Integration for Automotive (MEL), 2004

Collaborations and Standards Participation

Researchers in SIMA-supported projects work directly with 75 academic institutions; 36 consortia, national programs, professional and trade associations; and over 49 individual companies. Researchers in SIMA-supported projects are directly involved in the development of at least 38 formal interoperability standards. This work is conducted through over 50 organizations, including national and international formal standards organizations as well as various consortia and trade associations.

NIST Interoperability Testbed

Provides a venue for users and vendors to obtain information on integration specifications, and to test application interoperability during development and implementation. Users, especially small companies, are increasingly subject to costs and risks in the rapidly changing business environment. The testbed combines software testing expertise and domain expertise from multiple NIST labs. Using the testbed, specification developers receive feedback from software developers and users on implementation issues. Effective interoperability testing of approaches and facilities has proven to significantly reduce cost and cycle time associated with integration activities. The testbed will be offered as a web-accessible service, open to vendors, users, and researchers. The testbed will provide a principal mechanism for deployment of NIST integration research and infrastructure tools to industry.

Commonly used Acronyms

Commonly used Acronyms

A2LA	American Association for Laboratory Accreditation	ARL	Army Research Laboratory
AAAI	American Association for Artificial Intelligence	ARM	Application Reference Model
		ASA	Acoustical Society of America
AAMACS	Advanced Automated Master Angle Calibration System	ASACOS	Acoustical Society of America Committee on Standards
ACM	Association for Computing Machinery	ASCII	American Standard Code for Information Interchange
ADACS	Advanced Deburring and Chamfering System		
ADL	Architecture Design Language	ASME	American Society of Mechanical Engineers
AE	Acoustic Emission	ASPE	American Society of Precision Engineers
AE	Acoustical Emission	ASRS	Automated Storage and Retrieval Systems
AECMA	The European Association of Aerospace Industries	ASTM	American Society for Testing and Materials (now called ASTM)
AFM	Atomic Force Microscope, Atomic Force Microscopy	ATEP	Algorithm Testing and Evaluation Program
		ATM	Asynchronous Transfer Mode
AFMETCAL	Air Force Metrology and Calibration Program	ATP	Advanced Technology Program
AGA	American Gas Association	ATR	Advanced Technology and Research
AGMA	American Gear Manufacturers Association	ATS	Abstract Test Suite (related to STEP) or Algorithm Testing System
AGV Automated	GuidedVehicle		
AIAG	Automotive Industry Action Group	AWMS	Automated Welding Manufacturing System
AIM	Application Interpreted Model	AWS	American Welding Society
AIMS	Agile Infrastructure for Manufacturing Systems	B2B	Business-to-Business
AMSANT	Advanced Manufacturing Systems Applications Networked Testbed	B2B2C	Business-to-Business-to-Consumer
		B2C	Business-to-Consumer
AMT	Association for Manufacturing Technology	BCAC	Boeing Commercial Airplane Company
AMWA	Association for Metropolitan Water Agencies	BFRL	Building and Fire Research Laboratory
ANN	Artificial Neural Network	BIPM	Bureau International des Poids et Mesures (France)
ANSI	American National Standards Institute		
AP	Application Protocol	BITWO	Best in the World
APDE	Application Protocol Development Environment	BMP	Best Manufacturing Practices program
		CAD	Computer Aided Design
APEC	Asian Pacific Economic Cooperation	CAE	Computer Aided Engineering
API	American Petroleum Institute or Application Programming Interface	C-AFM	Calibrated Atomic Force Microscope
		CAID	Computer-Aided Industrial Design
APMP	Asia-Pacific Economic Cooperation	CALS	Continued Acquisition and Life Cycle Support

Commonly used Acronyms

CAM	Computer Aided Manufacturing	CNC	Computer Numerical Control
CAME	Computer Aided Manufacturing Engineering	COGM	Committee on Gear Metrology (ASME)
CAM-I	Consortium for Advanced Manufacturing - International	COM	Component Object Model
CAPP	Computer Aided Process Planning	COOMET	Euro-Asian Cooperation Of National Metrological Institutions
CASE	Computer-Aided Software Engineering	CORBA	Common Object Request Broker Architecture
CBM	Condition Based Maintenance or Condition Based Monitoring	CPM	Core Product Model
CBN	Cubic Boron Nitride	CRADA	Cooperative Research and Development Agreement
CC	Consultative Committee or common criteria	CSIRO	Commonwealth Scientific and Industrial Research Organization (Australia)
CCAUV	Comité Consultatif de l'Acoustique, des Ultrasons et des Vibrations / Consultative Committee for Acoustics, Ultrasound and Vibration	CSTL	Chemical Science and Technology Laboratory
		CXO	Chandra X-ray Observatory
CCD	Charge-Coupled Device	DAIS.	Data Acquisition from Industrial Systems
CCM	Consultative Committee on Mass and Related Quantities	DAML	DARPA Agent Markup Language
		DARPA	Defense Advanced Research Projects Agency
CD	Critical Dimension, Committee Draft, or Compact Disk	Dasa	DaimlerChrysler Aerospace
		DB	Database
CEN	European Committee for Standardization	DBMS	Database Management Systems
CENAM	Centro Nacional de Metrologia (Mexico)	DCOM	Distributed Component Object Model
CGPM	General Conference of Weights and Measures	DCS	Distributed Control Systems
CID	Charge Injection Device	DETC/CIE	Design Engineering Technical Conference/Computers in Engineering
CIFP	Capital Improvement of Facilities Project		
CIM	Computer Integrated Manufacturing	DFM	Danish Institute of Fundamental Metrology (Denmark)
CIMOSA	Computer Integrated Manufacturing Open Systems Architecture	DFM	Design For Manufacture
CIPM	International Committee of Weights and Measures (France)	DFT	Design for Tolerancing
		DHS	Department of Homeland Security
CIRP	College International Pour l'Etude des Techniques de Production Mecanique (International Organization for Production Engineering Research)	DIS	Draft International Standard
		DLL	Dynamic Link Libraries
		DMIS	Dimensional Measuring Interface Standard
CMM	Coordinate Measuring Machine	DMS	Distributed Manufacturing Simulation
CMP	Chemical-Mechanical Polishing	DMSO	Defense Modeling and Simulation Office
CMS	Coordinate Measuring System		

Commonly used Acronyms

DNC	Direct Numerical Control
DOC	Department of Commerce
DOD	Department of Defense
DOE	Department of Energy
DOT	DMIS Object Technology or Department of Transportation
DSP	Digital Signal Processors
DTD	Document Type Definition
DTM	Diamond Turning Machine
DTRA	Defense Threat Reduction Agency
DVM	Digital Voltmeter
E2M	Economically driven Environmentally conscious Manufacturing
EAL	European Cooperation for Accreditation of Laboratories
ECD	Electrical Critical Dimension
ECSS	Expert Control System Shell
EDI	Electronic Data Interchange
EDM	Electrical Discharge Machining
EEEL	Electronics and Electrical Engineering Laboratory
EIA	Electronic Industries Alliance (changed from Electronic Industries Alliance in 1997)
EIF	Enterprise Integration Framework
EMC	Enhanced Machine Controller
EMI	ElectroMagnetic Interference
EMMA	Easily Manipulated Mechanical Armature
EMS	Electronics Manufacturing Service
EPISTLE	European Process Industries STEP Technical Liaison Executive
EPRI	Electric Power Research Institute
ERM	Enterprise Resource Management
ERP	Enterprise Resource Planning
ETL	Electrotechnical Laboratory
EU	European Union
EUROMET	European Collaboration in Measurement Standards
EUSPEN	European Society for Precision Engineering and Nanotechnology
EUVL	Extreme Ultraviolet Lithography
FAMU	Florida Agricultural and Mechanical University
FBICS	Feature Based Inspection and Control System
FCIM	Flexible Computer Integrated Manufacturing
FDIS	Final Draft International Standard
FEA	Finite Element Analysis
FEM	Finite Element Modeling
FHWA	Federal Highway Administration
FIM	Field Ion Microscopy
FIPA	Foundation for Intelligent Physical Agents
FIPS	Federal Information Processing Standard
FLIR	Forward Looking Infra-Red
FMR	Field Material-handling Robot
FOT	Field Operations Test
FSU	Florida State University
FTD	Fabrication Technology Division
FTP	File Transfer Protocol
FY	Fiscal Year (October 1 - September 30)
GDP	Gross Domestic Product
GDRS	General Dynamics Robotic Systems
GD&T	Geometric and Dimensional Tolerancing
GERAM	General Enterprise Reference Architecture Model
GIF	Graphics Interchange Format
GMAW	Gass Metal Arc Welding
GOALI	Grant Opportunities for Academic Liaison with Industry (NSF program)

Commonly used Acronyms

GPL	General Purpose Laboratory
GPS	Global Positioning System or Geometric Product Specification
GUI	Graphical User Interface
GUM	ISO Guide to the Expression of Uncertainty in Measurements
HLA	High Level Architecture
HMMWV	High Mobility Multipurpose Wheeled Vehicle
HPCC	High Performance Computing and Communications
HRC	Rockwell C Hardness
HTML	HyperText Markup Language
HTTP	Hyper Text Transfer Protocol
IAI	Industry Alliance for Interoperability
IAMS	Institute of Advanced Manufacturing Sciences
IAV	Industrial Autonomous Vehicle
IBIS	Integrated Ballistics Identification System
IC	Intelligent Control, International Comparison, or Integrated Circuit
ICM	Internet Commerce for Manufacturing
IDEF	Integrated Computer-Aided Manufacturing Definition
IDL	Interface Definition Language
IE	Industrial Engineering, Invested Equipment
IEC	International Electrotechnical Commission
IEEE	Institute of Electrical and Electronic Engineers
IFIP	International Federation for Information Processing
IGES	Initial Graphics Exchange Specification
IGRIP	Interactive Graphics Robot Instruction Program
IGT	Institute for Gas Technology
IITA	Information Infrastructure Technology Applications (HPCC program)
IMES	Initial Manufacturing Exchange Specification
IMES	Initial Manufacturing Exchange Specification
IMGC	Instituto di Metrologia "G. Colonnetti" (Italy)
IMS	Intelligent Manufacturing Systems
IMTI	Integrated Manufacturing Technology Initiative
IMTR	Integrated Manufacturing Technology Roadmap (http://imtr.ornl.gov/)
IMU	Inertial Measurement Unit
INAOE	Instituto Nacional de Astrofisica, Optica, y Electrónica in Tonantzinla, Puebla, Mexico
INMETRO	Instituto Nacional de Metrologia, Normalizacao e Qualidade Industrial (Brazil) Brazilian National Measurement Institute in Rio de Janeiro
INS	Inertial Navigation System
IP	Internet Protocol
IPM	Intelligent Processing of Materials
IPO	IGES/PDES Organization
IPR	In Process Review
IR	Infrared (light)
IS	International Standard
ISAM	Indexed Sequential-Access Management/Method
ISATP	International Symposium on Assembly and Task Planning
ISD	Intelligent Machines Division
ISIC/CIRA/ISAS	Three combined conferences: IEEE International Symposium on Intelligent Control, IEEE International Symposium on Computational Intelligence in Robotics and Automation, and the Intelligent Systems and Semiotics conference
ISMT	International SEMATECH
ISO	International Organization for Standardization – not an acronym according to ISO
ISOMA	International Symposium on Manufacturing and Applications
ISPE	International Society for Pharmaceutical Engineering

175

Commonly used Acronyms

IT	Information Technology or Industrial Technology
ITA	Interim Testing Artifact
ITL	Information Technology Laboratory
ITRS	International Technology Roadmap for Semiconductors
ITS	Intelligent Transportation Systems
ITU	International Telecommunication Union
JAST	Joint Advance Strike Technology (a DOD program)
JAUGS	Joint Architecture for Unmanned Ground Vehicle Systems (DOD)
JCGM	Joint Committee for Guides in Metrology
JEDMICS	Joint Engineering Data Management Information Computer Systems
JPEG	Joint Photographic Experts Group
JPL	Jet Propulsion Laboratory
JSW	Joint Standards Workshop
KEMAR	Knowles Electronics Manikin for Acoustic Research
KIC	Key International Comparison
KIF	Knowledge Interchange Format
KRISS	Korea Research Institute of Standards and Science
LARCS	Laser Rail Calibration System
LDWS	Lane Departure Warning Systems
LFAD	Laser-Focused Atomically Deposited
LIGO	Laser Interferometric Gravitational-wave Observatory
LLNL	Lawrence Livermore National Laboratory
LM	Layered Manufacturing
LMT	Large Millimeter Telescope
LSI	Line Scale Interferometer
LVDT	Linear Variable Differential Transformer
M2A	Manipulation, Measurement, and Assembly
M3	Molecular Measuring Machine
MAA	Metrology Automation Association
MADE	Manufacturing Automation Design Engineering Project (a DARPA project)
MANTECH	Manufacturing Technology (a DOD program)
MAP	Measurement Assurance Program
MBE	Molecular Beam Epitaxy
MCITT	Manufacturer's CORBA Interfact Testing Toolkit
MEL	Manufacturing Engineering Laboratory
MELSA	MEL System Administration team
MEMS	Micro-electromechanical systems
MEP	Manufacturing Extension Partnership
MES	Manufacturing Execution System
METK	Manufacturing Engineering Tool Kit
MfgDTF	Manufacturing Domain Task Force (OMG)
MfgTF	Manufacturing Task Force (an OMG program)
MFPT	Society for Machinery Failure Prevention Technology
MIDMAP	Mid-America Measurement Assurance Program — regional grouping of U.S. state metrology labs for national key comparisons
MISSION	Modeling and Simulation Environments for Design, Planning and Operation of Globally Distributed Enterprises
MITT	Manufacturing Information Technology Transfer
MOB	Mobile Offshore Base
ModSAF	Modular Semi-Automated Forces, a standard military simulation tool
MPA NRW	Materialprüfungsamt Nordrhein–Westfalen (Germany)
MR	Manufacturing Resource
MRD	Materials Reliability Division (NIST Materials Science and Engineering Laboratory)

Commonly used Acronyms

MRP	Material Requirements/Resource Planning	NII	National Information Infrastructure
MSA	Microscopy Society of America	NIIIP	National Industrial Information Infrastructure Protocols
MSEL	Materials Science and Engineering Laboratory	NIJ	National Institute of Justice
MSI	Manufacturing Systems Integration	NIM	National Institute of Metrology (China)
MSID	Manufacturing Systems Integration Division	NIST	National Institute of Standards and Technology
MTBF	Mean Time Between Failures	NISTIR	National Institute of Standards and Technology Interagency/Internal Report
MWG	Technical Working Group under SIM		
NADC	Naval Air Development Center	NIST-IR	NIST Industrial Robot
NAMAS	National Measurement Accreditation Service (England)	NITS	NIST - ITI Test System
		NMI	National Metrology (Measurement) Institute
NAMT	National Advanced Manufacturing Testbed	NORAMET	North American Metrology Cooperation (with NRC of Canada and CENAM of Mexico)
NASA	National Aeronautics and Space Administration		
		NPL	National Physical Laboratory (U.K. or India)
NSA	National Security Agency	NRC	National Research Council (U.S. and Canada)
NAVLAP	National Voluntary Laboratory Accreditation Program	NRL	Naval Research Laboratory
NC	Numerically controlled (machine tools and equipment)	NRLM	National Research Laboratory of Metrology (Japan)
NCAP	Network Capable Application Processor	NRO	National Reconnaissance Office
NCMS	National Center for Manufacturing Sciences	NSA	National Security Agency
NCS A&T	North Carolina State Agricultural and Technical University	NSF	National Science Foundation
		NSRP	National Shipbuilding Research Program
NCSL	National Conference of Standards Laboratories	NTA	National Technical Association
NDE	Non-Destructive Examination	NTEP	National Type Evaluation Program
NEMAP	Northeast Measurement Assurance Program — regional grouping of U.S. state metrology labs for national key comparisons	NTRS	National Technology Roadmap for Semiconductors
NEMS	nanoelectromechanical systems	NVLAP	National Voluntary Laboratory Accreditation Program
NGIS	Next Generation Inspection System	OA	Other (government) Agency
NGM	Next Generation Manufacturing	OAC	Open Architecture Control
NHTSA	National Highway Traffic Safety Administration	OAG	Open Applications Group
NIAP	NIST/NSA National Information Assurance Partnership	OAGIS	Open Applications Group Integration Specification
NICS	NIST Identifier Collaboration Service	OAS	Organization of American States
NIF	National Ignition Facility		

Commonly used Acronyms

Acronym	Definition
OASIS	Organization for the Advancement of Structured Information Standards
OEM	Original Equipment Manufacturer
OEOSC	Optics and Electro-Optics Standards Council
OI	Operator Interface
OIDA	Optoelectronics Industry Development Association
OIML	Organization Internationale de Metrologie Legale International Organization of Legal Metrology
OLES	Office of Law Enforcement Standards (NIST)
OMA	Object Management Architecture
OMAC	Open Modular Architecture Controller
OMG	Object Management Group
OMP	Office of Microelectronics Programs
OpenADE	Open Assembly Design Environment
ORB	Object Request Broker
ORMC	Oak Ridge Metrology Center
ORNL	Oak ridge National Laboratory
OSD	Office of the Secretary of Defense
OSRM	Office of Standard Reference Materials
OWL	Web Ontology Language
OWM	Office of Weights and Measures (NIST)
PAC	Personal Air Conditioning
PC	Printed Circuit or Personal Computer
PCMCIA	Personal Computer Memory Card International Association
PCSRF	Process Control Security Requirements Forum
PDES	Product Data Exchange Using STEP
PDM	Product Data Management
PED	Precision Engineering Division
PKM	Parallel Kinematic Machines
PL	Physics Laboratory
PLC	Programmable Logic Controllers
PLM	Polarized Light Microscopy
PMI	Phase Measuring Interferometer
POAC	Precision Optoelectronics Assembly Consortium
POSC	Petrotechnical Open Software Corporation
PRT	Platinum Resistance Thermometer
PSD	Power Spectral Density
PSL	Process Specification Language or Problem Statement Language
PTB	Physikalisch-Technische Bundesanstalt (Germany)
Pt-Ir	Platinum Iridium
PVDF	Polyvinylidenefluoride
PWI	Preliminary Work Item
PZT	A ceramic piezo-electric material used in actuators and ultrasonic
Q&V	Qualification and Validation
QIA	Quality In Automation
R&D	Research and Development
RaDEO	Rapid Design Exploration and Optimization program
RAMM	Rapid Agile Metrology for Manufacturing (an ATP sponsored program)
RAMP	Rapid Acquisition of Manufactured Parts (a DoD sponsored program)
RCS	Real-Time Control System
RCSA	Receptance Coupling Substructure Analysis
RDF	Resource Description Framework
RFP	Request For Proposals
RIA	Robotics Industries Association
RM	Research Material
RMS	Root Mean Square
ROI	Return On Investment
RP	Rapid Prototyping
RRL	Rapidly Renewable Lap

Commonly used Acronyms

RRM	Rapid Response Manufacturing
RSS	Root Sum Squares
RST	Robot Systems Technology
RSTA	Reconnaissance, Surveillance, and Target Acquisition
RTI	Run-Time Infrastructure or Real Time Innovations, Inc.
SAE	Society of Automotive Engineers
SAIC	Science Applications International Corporation
SBIR	Small Business Innovation Research
SC	Subcommittee (under ISO)
SCADA	Supervisory Control and Data Acquisition
SCLD	Scanning Capacitance Line Detector
SDAI	Standard Data Access Interface
SDX	Simulation Data Exchange
SEI	Software Engineering Institute at Carnegie Mellon University
SEM	Scanning Electron Microscope, Scanning Electron Microscopy
SEMAP	Southeast Measurement Assurance Program regional grouping of U.S. state metrology labs for national key comparisons
SEMATECH	SEmiconductor MAnufacturing TECHnology consortium (formally changed to International SEMATECH in 2000)
SEMI	Semiconductor Equipment and Materials International
SET	Single Electron Tunneling
SFF	Solid Freeform Fabrication
SGML	Standard Generalized Mark-up Language (an ISO standard)
SI	Systeme Internationale d'Unites (the modern metric system)
SIA	Semiconductor Industry Association
SIF	Solid Interchange Format
SIGMA	Supersonic Inert Gas-Metal Atomizer
SIM	Systema Interamericano de Metrologia; Interamerican System of Metrology (English translation)
SIMA	Systems Integration for Manufacturing Applications
SIMOX	Separation by Implantation of Oxygen
SIP	STEP Implementation Prototype
SLA	Stereolithography apparatus
SME	Small/Medium Enterprise
SOI	Silicon-On-Insulator
SOLIS	SC4 On-Line Information Services
SPIE	The International Society for Optical Engineering
SPM	Scanning Probe Microscopy, Scanning Probe Microscope
SQL	Standard Query Language
SRM	Standard Reference Material
STEP	STandard for the Exchange of Product model data
STL	Stereolithography
STM	Scanning Tunneling Microscope, Scanning Tunneling Microscopy
STRS	Scientific and Technical Research and Services (NIST appropriated budget)
STS	Scanning Tunneling Spectroscopy
SWAP	Southwest Measurement Assurance Program — regional grouping of U.S. state metrology labs for national key comparisons
SXPL	Soft X-ray Projection Lithography
TACS	Tactical Auxiliary Crane Ship
TAG	Technical Advisory Group
TC	Technical Committee (under ISO)
TCP	Transmission Control Protocol
TDP	Technical Development Plan, Technical Data Package
TEAM	Technologies Enabling Agile Manufacturing (DOE)

Commonly used Acronyms

TEDS	Transducer Electronic Data Sheet	WRAP	West Region Measurement Assurance Program — regional grouping of U.S. state metrology labs for national key comparisons
TEM	Transmission Electron Microscope, Transmission Electron Microscopy	WWW	World Wide Web
TIDE	Technology Insertion, Demonstration, and Evaluation program of SEI	XCALIBIR	X-ray Optics Calibration Interferometer
TIMA	Technologies for the Integration of Manufacturing Applications (ATP program)	XML	Extensible Markup Language
		XRT	X-ray Topography
TIMS	Testing of Interaction-driven Manufacturing Systems	XSLT	eXtensible Style Language for Transformations
TIS	Tool Induced Shift	XUV	Next Generation Unmanned Vehicle
TLM	Technology Learning Modules		
UdeBW	Universitat de Bundeswehr (German)		
UGV	Unmanned GroundVehicle		
UHV	Ultrahigh Vacuum		
ULSI	Ultra Large Scale Integration		
UML	Universal Modeling Language		
UN/CEFACT	United Nations Centre for Trade Facilitation and Electronic Business		
USNC	United States National Committee of IEC		
USNWG	United States National Working Group		
US TAG	United States Technical Advisory Group		
USPRO	United States Product Data Association		
UTAP	Unified Telerobotic Architecture Project		
UTRC	United Technologies Research Center		
UV	Ultraviolet (light)		
VA	Department of Veterans Affairs		
VIM	International Vocabulary of Basic and General Terms in Metrology		
VLSI	Very-Large-Scale Integrated...		
VRML	Virtual Reality Modeling Language		
W	Tungsten		
W&M	Weights & Measures		
WG	Working Group		
WGDM	Working Group on Dimensional Metrology		

Commonly used Acronyms

www.ingramcontent.com/pod-product-compliance
Lightning Source LLC
Chambersburg PA
CBHW081723170526
45167CB00009B/3679